AF602045

Ecology of Similipal Biosphere Reserve

The Authors

Dr. Rabindra Kumar Mishra is presently working as Lecturer in the Department of Wildlife and Biodiversity Conservation, North Orissa University, Takatpur, Baripada. He received his under graduate, Post-graduate and Ph. D. Degree from Utkal University, Vani Vihar, Bhubaneswar. He has teaching and research experience in the field of Ecology especially in Forest Ecology. He has published over 40 research papers in the journals of International and National repute having impact factors. Odisha Bigyan Academy Young Scientist Award was awarded to him in 2007 for his research contribution in Life Science particularly in phytosociology. He was also one of the recipients of Young Scientist Fellow of Department of Science and Technology, Govt. of India, New Delhi from 2005 to 2008. Dr. Mishra has participated and delivered lectures at various National and International seminars and symposia. He is also a member of several scientific societies. He was selected as one of the Functional Area Expert (FAE) in the field of Ecology and Biodiversity, and Soil Conservation by Quality Council of India (QCI), Govt. of India, New Delhi.

Dr. Anup Kumar Nayak received M. Sc. from Utkal University and AIFC Diploma in Forestry from Indian Forest College, Dehradun. An Indian Forest Service Officer of 1985 batch, he obtained Ph. D. in Avifauna of Bhitarkanika from Utkal University in 2009. He has been working in the field of Wildlife Conservation for the last 14 years since his tenure as Divisional Forest Officer (DFO), Chilika in 2000. He has written many articles in national and international journal on birds, mangroves and sea turtles. He is an active member of IUCN Species Survival Commission for threatened waterfowl subgroup since 2006. He was a Professor and Head of Department of Landscape Level Planning and Management in Wildlife Inststute of India, Dehradun from 2006-2009. He has also worked as Joint Director in Wildlife Crime Control Bureau, Ministry of Environment and Forests, New Delhi of two and half years from 2009-2011 and Field Director, Similipal Biosphere Reserve-cum- Regional Chief Conservator of Forests, Baripada, Odisha from 2011-2014. Currently he works as Regional Chief Conservator of Forests of Angul district of Odisha.

Dr. Hemanta Kumar Sahu has passed M. Sc. and M. Phil from Utkal University, Odisha. He was also awarded Ph. D. from the same University in 2000. He has done his Ph. D. in ducks and geese of Chilika lake. He has also served in Wildlife Institute of India, Dehradun as Project Associate and Faculty in Wildlife and Biodiversity Conservation of North Odisha University in 2002. Now he is working as Reader, Department of Zoology, North Orissa University. He is involved in Wildlife Research in Odisha as well as in North Eastern States of India and published papers on wildlife of those areas. He has also visited Smithsonian Institute, USA on training programmes for Biodiversity and Statistical Course during 2006, 2010 and 2014.

Himanshu Shekhar Palei is an alumnus of Department of Wildlife and Conservation Biology and doing Ph.D. under the guidance of Dr. H.K. Sahu at Department of Zoology, North Orissa University. Himanshu has research interests in the ecology and conservation of threatened Wildlife and aims to develop regional conservation strategies for threatened wildlife of Odisha. He is currently working at Wildlife Institute of India, Dehradun and investigating how the dynamics of threatened wildlife populations are influenced by mining in Odisha.

Ecology of Similipal Biosphere Reserve

– Authors –
Rabindra Kumar Mishra
Anup Kumar Nayak
Hemanta Kumar Sahu
Himanshu Shekhar Palei

2016
Daya Publishing House®
A Division of
Astral International Pvt. Ltd.
New Delhi - 110 002

Publisher's note:

Every possible effort has been made to ensure that the information contained in this book is accurate at the time of going to press, and the publisher and author cannot accept responsibility for any errors or omissions, however caused. No responsibility for loss or damage occasioned to any person acting, or refraining from action, as a result of the material in this publication can be accepted by the editor, the publisher or the author. The Publisher is not associated with any product or vendor mentioned in the book. The contents of this work are intended to further general scientific research, understanding and discussion only. Readers should consult with a specialist where appropriate.

Every effort has been made to trace the owners of copyright material used in this book, if any. The author and the publisher will be grateful for any omission brought to their notice for acknowledgement in the future editions of the book.

Cataloging in Publication Data--DK
Courtesy: D.K. Agencies (P) Ltd. <docinfo@dkagencies.com>

Mishra, Rabindra Kumar *(Lecturer in Wildlife and Biodiversity Conservation)*, author.
Ecology of Similipal Biosphere Reserve / authors, Rabindra Kumar Mishra, Anup Kumar Nayak, Hemanta Kumar Sahu, Himanshu Shekhar Palei.
pages cm
Includes bibliographical references.
ISBN 9789390384136 (Hardbound)

1. Forest ecology--India--Similipal Biosphere Reserve. 2. Forest biodiversity conservation--India--Similipal Biosphere Reserve. 3. Forest policy--India--Similipal Biosphere Reserve. I. Nayak, Anup Kumar, 1960- author. II. Sāhu, Hemanta Kumāra, author. III. Palei, Himanshu Shekhar, author. IV. Title.

QK938.F6M57 2016 DDC 577.30954 23

Published by : **Daya Publishing House®**
A Division of
Astral International Pvt. Ltd.
– ISO 9001:2008 Certified Company –
4760-61/23, Ansari Road, Darya Ganj
New Delhi-110 002
Ph. 011-43549197, 23278134
E-mail: info@astralint.com
Website: www.astralint.com

Laser Typesetting: SSMG Computer Graphics, Delhi - 110 084

Printed at : Replika Press Pvt. Ltd.

NORTH ORISSA UNIVERSITY
Sriram Chandra Vihar, Takatpur,
Baripada, Mayurbhanj -757 003, Odisha

Phone : 06792-255127 (O), 278275 (R)
Cell : 94375 70288, Fax : 06792-253908
E-mail : vconou@rediffmail.com

Prof. Prafulla Kumar Mishra
Vice-Chancellor

Foreword

Similipal is a range of mountain like other kutacalan i.e. Vindhya, Mahendra, Sahya and Aravali, etc. Located in Mayrbhanj district of Odisha in the eastern part of India, representing a biological wealth of option for the present and future generations and is considered as the Himalaya of Odisha. Because of its unique floral, faunal and ecosystem diversity, it is included in the UNESCO's list of few international biosphere reserves of our country. It is also equally rich in traditional knowledge for conserving the resources of various kinds including biodiversity. In the recent past several attempts have been made to document its rich biological resources. However, studies related to ecology of plants and animals of the reserve, essential for its biodiversity conservation is still lacking and yet to be explained.

The use of ecological principles in addressing the biodiversity conservation issues in increasingly gaining ground among the practitioners. In fact, in order to have sustainable long-term solution to this issue, solution based on sound ecological principles seem to be the only answer. With the changing world, where the components of biodiversity are increasingly threatened due to deforestation, habitat fragmentation and human activities, ecological studies like phytosociology and phenology of various plant life forms (herbs, shrubs, trees and climbers), litter decomposition and nutrient turn over on the forest floor, dispersion of plants and categorization of animals as per IUCN categories providing scientific inputs for biodiversity conservation.

It is indeed a matter of great pleasure for me to known that this publication provides information on various threats to the biodiversity of Similipal on one hand and proper use of the results of ecological studies on the other hand for better management of its floral and faunal resources. This publication will be immensely useful for a wide array of people including general public, academicians and students gaining knowledge with an interest on plant and animal life in general and protected area mangers, conservationists, research scholars and administrators on formulating management plans for conservation of biological resources in particular.

(Prafulla Kumar Mishra)

Preface

Similipal is one of the premier tiger reserve of the country and the only biosphere reserve of Odisha. The diverse topographic features, wide altitudinal variations, varied micro climatic conditions, unique geographical position and a wide range of forest ecosystems have contributed to its unique floral and faunal diversity. The area is inhabited by various tribal communities are rich in their traditional knowledge for conserving various natural resources including the biodiversity of Similipal. However, of late activities like deforestation, habitat fragmentation, forest fire, grazing and other anthropogenic activities have posed serious threats to its unique biodiversity. Though it has occupied a special position not only in the biodiversity map of India but also in the world still then its biodiversity is in the state of degradation due to the nomadic and semi-nomadic activity of foragers (forest dwellers) and lack of proper conservation measures. To effectively address the biodiversity conservation issues of Similipal is not an easy task because of the complexities attached over it with respect to social, political and scientific background.

The scientific role especially of ecological principles with respect to the issues related to biodiversity conservation of Similipal is increasingly gaining ground among the practitioners, planners and the forest managers. In fact, in order to have a sustainable long-term solution to this issue, application of sound ecological principles seem to be the only answer. Therefore scientific inputs from ecological studies on forest ecosystems of Similipal with other allied disciplines are emphasized here. The book entitled "Ecology of Similipal Biosphere Reserve" is the outcome of information gathered on ecological studies and allied disciplines carried out in Similipal Biosphere Reserve (SBR) from time to time. The book has been divided into eight chapters viz. (i) the forest types and dynamics, (ii) structure, composition and phytosociology of plant life forms, (iii) litter accumulation and decomposition on forest floor, (iv) phenology of upperstorey, middle storey and understorey or ground flora (v) herpetofaunal diversity, (vi) bird diversity, (vii) mammal diversity and (viii) deforestation and degradation of forest covers of Similipal. Presentation of each chapter in this volume is self explanatory and has not only been evoked out of a mere academic interest, but also used in the long run towards better management of the natural resources of Similipal.

Rabindra Kumar Mishra
Anup Kumar Nayak
Hemanta Kumar Sahu
Himanshu Shekhar Palei

Contents

	Foreword	*v*
	Preface	*vii*
1.	**Introduction**	**1**
2.	**Study Area**	**5**
	Topography	7
	Water Resource of Similipal	8
	Climate	8
	Geology and Soil	8
	Human Habitations	10
	Road Connections	10
	Mineral Deposit of SBR	10
3.	**Forest types and dynamics of Similipal**	**11**
	Northern Tropical Moist Deciduous Forests	12
	Northern Tropical Semi-Evergreen Forest	14
	Dry Deciduous Hill Forests	14
	High Level Sal Forests	15
	Low Level Sal Forest	16
	Grassland and Savannah	16
	Plain Sal Forest	17
	Dry Sal Forests	18

	Forest Dynamics	19
	Structural Dynamics of Tropical Forests of Similipal	21
	Ecological Dynamics	21
	Change in Species Richness (Species Dynamics)	22
	Structural Dynamics	23
4.	**Structure, Composition and Phytosociology of Plants**	**32**
	Methods of Study of Phytosociology of Plants	34
	Floristic Compositions and Occurence	34
	Raunkiaer's Classification of Vegetation Layers of SBR	35
	Distribution Pattern	35
	Distribution of Climbers	52
	Ecological Importance of Species	53
	Stand Structure	53
	Diversity Measures	55
	Comparative Analysis of Tree Species Diversity Under Various Tropical Forests	56
5.	**Litter Decomposition as an Ecosystem Service**	**59**
	Methods to Study Litter Decomposition	60
	Plant Diversity and Litter Depth	61
	Forest Floor Biomass	62
	Rate and Time of Forest Floor Biomass	66
	Factors Affecting Litter Decomposition	68
	Climatic Factors	69
	Effects of Vegetation on Forest Floor Biomass	70
	Forest Floor Biomass and Soil Physical Characters	71
	Forest Floor Biomass and Soil Chemical Characters	75
	Soil Biota	75
	Relationship Between Abiotic Factors and Forest Floor Litter Decomposition and Turn Over	75
	Patterns of Litter Decomposition and Nutrient Release	76
6.	**Phenology of Plant Life Forms of Similipal**	**78**
	Sampling and Collection of Phonological Data	79
	Duration and Pattern of Activity	79
	Fruit Maturation Activity	80
	Phenophases of Overstorey and Understorey Species of Similipal	80

Flowering Activity 91
Fruiting Activity 92
Ranges of Interphenophase Durations 93
Phenophases of Ground Flora 93
Future Activities on Phenophases of Plant Life Forms 94

7. Herpetofaunal Diversity of Similipal 95
Herpetological Inventories 96
Herpetofauna Diversity of Similipal Biosphere Reserve 96
Conservation Measures 97

8. Status of Birds in Similipal 102
Ornithological History of Similipal 103
Composition of Birds of Similipal 103
Globally Threatened Species 104
Water Birds and Winter Migrants 104
Conservation Issues and Implications 104

9. Status of Mammals in Similipal 127
Methods of Study of Mammals 129
Composition of Mammals 129
Implications for Conservation 131
Relative Abundance Index 135

10. Deforestation and Degradation of Forest Covers in Similipal 136
Illegal Cutting of Trees 137
Illegal Removal of NTFP 140
Encroachment and Other Illegal Activities 142
Human Impact on Similipal Forests 148
Domestic Livestock Grazing 148
Wild Fire 149
Insect Attacks and Pathological Problems 152
Soil Nutrient Status and Degradation 152
The Illegal and Unscientific Collection of Medicinal Plants 154
Poaching of Faunal Species 154

***References* 156**
***Index* 176**

Chapter-1

INTRODUCTION

Forest ecosystem is one of the complex systems in which plants and animals not only interact among themselves but also interact with the abiotic factors to maintain the biological diversity. The biological diversity is a dynamic feature that provides resistance and stability, taste and savior, colour and beauty, vigour and validity to a system. It also bestows similar assets to ecological system in our environment. For conservation of the biological diversity, the network of national parks, sanctuaries, biosphere reserves and other protected areas should be strengthened and extended adequately. Biosphere reserves are one of the natural Protected Areas (PAs) included in a global network organized by the United Nations Educational, Scientific, and Cultural Organization (UNESCO) play a critical role in protecting our natural heritage. They are set up to preserve unique and rare plants, animals and geological features. Our goal is to protect or safeguard them from all sorts of disturbances. The Biosphere Reserves represent characteristic ecosystems in different biogeographic regions and consider human communities as their integral component. In a broader sense the biosphere reserve ensuring *in-situ* conservation at all levels of biodiversity ranging from genes to ecosystems, widening the understanding of components of ecosystems through research and monitoring and achieving integrated development of the area. The BRs are, therefore, sites for experimenting with learning about sustainable development (UNESCO, 2011). The biosphere reserves provide opportunities for people to enjoy the educational and aesthetic benefits of natural areas. They are not formed for outdoor recreation and should not be confused with parks or other recreational areas. The creation of biosphere reserves is an integral part of society's efforts to maintain biodiversity. Biodiversity refers to the complex web of life including all species from microscopic bacteria to plants and animals. They afford us opportunities to study the biodiversity of the province and provide benchmarks against which environmental changes can be measured, particularly where species are growing at the limits of their range. Biosphere reserves are also genetic data banks which may hold the key to new discoveries in forestry, ecology, agriculture and medicine.

The Indian National Man and Biosphere Committee identify and recommend potential sites for designation as Biosphere Reserves, following the UNESCO's guidelines and criteria. The Biosphere Reserves are different from wildlife sanctuaries and National Parks as the emphasis is on overall biodiversity and landscape rather than on specific species. In a biosphere, there are three different zones: core, buffer and transition zones. The core areas are the most heavily protected sites where the ecosystems remain relatively untouched. They are the areas designated mainly for conservation, and the only human role in these areas is for observation and non destructive research. The buffer zones surround the core area are open to people to visit. However, people cannot inhabit these areas. They are mainly used for recreation and ecotourism. Transition zones are the areas in the biosphere where people can inhabit. The people living in these areas are usually scientists, management agencies, and cultural groups native to the area and farms and fisheries (biosphere). The people who live in this area are usually responsible for managing, sustaining, and developing the biosphere.

Odisha is one of the state quite rich in natural resources and has several areas very much similar to biodiversity hotspot areas. It has varied and widespread forest covers in different parts viz. dry deciduous, moist deciduous forests as well as mangroves with several unique, endemic, rare and endangered floral and faunal species. Odisha ranks fifth amongst State/Union Territories of the country in terms of area under forest cover. The total forest area of the State is 58, 136 sq km, which is 31.4 per cent of the State's geographical area and about 7.66 per cent of the country's forests (FSI, 2011).

Forest ecosystem denotes the diversity of habitats, communities and ecological processes within the biosphere. Studies of forest ecosystem diversity are carried out at different scales, from one forest ecosystem to an entire region containing several different forest ecosystems. Regions containing a great variety of ecosystems are rich in biodiversity. Similipal in Odisha located in the central part of the Mayurbhanj district of Odisha, with a total area of 5569 km^2 is such an area where diverse ecosystem types rich in species and genetic diversity are noticed. Therefore this region is one of the richest zones of biodiversity in Odisha as well as in India. It is separated from the Bay of Bengal by a narrow strip of coastal plain and contains only 38% of the total protected area network of Odisha but its diverse topographic feature, wide altitudinal variation, varied climatic condition, unique geographical location and wide range of ecosystem diversity has contributed to the rich and unique floral and faunal diversity. It is the home to about 1076 species of plants, 94 species of orchids, 42 species of major mammals, 264 species of birds, 12 species of amphibians and 30 species of reptiles and considered as the Himalaya of Odisha (Biswal et al., 2011). In addition to the outstanding levels of species diversity and endemism, the eco-region also plays an important role in maintaining altitudinal connectivity between the habitat types. The topography, the altitudinal variation and the peaks along with forest wilderness have enriched the area with biodiversity. It is also the richest watershed in Odisha, giving rise to many perennial rivers viz. East Deo,West Deo, Khair, Bhandan, Budhabalang, Sanjo, Salandi, Palpala, etc. Gorgeous Barehipani and Joranda waterfalls are of great attractions. The Barehipani waterfall is located at the centre of the area. It is one of the tallest waterfalls in the country, at a height of 399 m. The falls are the most beautiful sites of the area to attract the tourists and also the perennial water source of rivers of this region. Based on the rich biodiversity and the unique physical and topographical features it was declared Odisha's 1st and country's 8th biosphere reserve on 22nd June, 1994 under UNESCO's Man and Biosphere programme. However, of late, deforestation, live stock grazing, fragmentation or formation of micro habitats, forest fire, and conversion of forest lands into agricultural lands and extraction of forest products in an unsustainable way and other human activities have posed serious threat to the rich biodiversity of Similipal. Livestock grazing inside SBR causes competitive exclusion of wild animals from high-quality habitats because they may be forced to forage in poor habitats and spend more energy to move away from the disturbances that affect their nutritional balance (Schaller, 1977). Ungulates are a major constituent of mammalian fauna of Similipal and also the major prey base for large mammalian predators. They are known to modify their activity pattern in response to habitat differences, seasons, and disturbance factors and their behaviour could be a sensitive indicator signifying the habitat quality.

Fragmentation due to burgeoning human population and agricultural expansion both in transitional and buffer areas of the reserve leads to conversion of the natural forest covers into smaller and non-contiguous patches, is the most serious threat to the long-term survival of the biological diversity of Similipal. The fragmented patches in some pockets of SBR embedded in a matrix of anthropogenically manipulated landscapes behave as islands in a sea of pasture or agricultural ecosystem and may lead to distinct ecological, demographic and genetic consequences result in the extinction of native species. Fragmentation or conversion of forest into grass land or savanna due to forest harvesting, fertilization and change in abiotic factors also affect the nitrogen mineralization in different forest types of Similipal (Tripathi and Singh, 2012). Forest fragmentation is not only affected the plants but also the large predators that play a key role in regulating herbivore or prey populations (Duffy, 2003). In addition to this fragmentation caused due to logging practice also affects the community dynamics of the over storey trees by creating gaps with altered light and microclimate conditions, facilitating the growth of small-seeded, light loving pioneer species (Pelissier et al., 1998). Intensive logging of over storey species by the wood smugglers and fringe area villagers reduce populations of old growth species. As these species are slow-growing, they may take many years to recover their population in the natural forest covers is drastically reduced. Many of them are important fruiting resources for animals, which in turn disperse the seeds of these species. Logging may thus reduce overall densities and alter relative abundances of frugivorous dispersers either directly or indirectly through loss of food trees (Cleary et al., 2007; Sethi and Howe, 2009). Loss of dispersers may eventually result in reduction or loss of biotically-dispersed trees and change the functional composition of the forest towards abiotically-dispersed species.

The ecological consequences of forest fragmentation on herbivore, carnivore, bird and other types of animal population of Similipal are still untouched, which needs the attention of the biosphere reserve Manager. Information on comprehensive census of the flora and fauna, detection of food habit of herbivores, identification of fodders, evaluation of mycorrhizal associations and other symbionts, detection of soil profile, detection of resources for water and salt supplementations, selection of plant propagules and determination of plantation method are also important with respect to conservation of habitats and wild lives of SBR. Evaluation of tree species in climax vegetation is very important because a tree species creates an ecological niche of its own (Stork, 1991; Andrea and Russell, 2005) and denudation of it may bring forth a cascade effect resulting in the defaunation of a biodiversity (Leigh et al., 1993). Considering the above facts the scope of this book to gather information and discuss about (i), the forest types and dynamics, (ii) structure, composition and phytosociology of plant life forms, (iii) litter decomposition on forest floor, (iv) phenology of upperstorey, middle storey and understorey or ground flora, (v) herpeto faunal diversity, (vi) bird diversity, (vii) mammal diversity and (viii) deforestation and degradation of forest covers of Similipal.

Chapter-2

STUDY AREA

Similipal forests came into lime light when it was brought under the fold of management through a Forest Policy formulated by the then Maharaja of Mayurbhanj in the year 1885. The wildlife in Similipal Forests was managed primarily for recreation of the Royal family and their guests. 'Akhand Shikar' (tribal hunt) was celebrated continuously for five to seven days during the month of April every year. The forest was worked as per the Working Plan prepared by Mr. C.C. Hart in 1896-97 till 1946, through lease and contract. The main purpose of timber harvesting was to earn revenue for the Royal exchequer. Since 1946, the forest was exploited systematically as per prescription of Working Plan till merger of the ex-state in the Union of India on 6th November 1948, which became part of Odisha a district on 1st January 1949. Despite practising commercial forestry, supplying railway sleepers and other utility timber outside Mayurbhanj, the ruler was very rigid in his forest protection measures and employed large number of forest staff, much higher in number in comparison to other princely states and even the directly British administered areas with good network of forest roads and communication facilities. The protection of forest suffered a setback after independence due to issue of liberal shooting permits. Although there was depletion of wildlife still the number of tigers inhabiting the area was potentially high. The era of protection of forestry started for the wildlife after declaration of Similipal Tiger Reserve during the year 1973 with Shri S.R. Choudhury as the first Field Director as per policy of Govt. of India under "Project Tiger". Similipal Tiger Reserve (STR) created with the adjoining forest blocks of Similipal RF extending over 2750 km^2. Then Similipal reserved forest was notified as proposed sanctuary on 3rd December 1979 which was followed in quick succession on 6th August 1980 and 11th June 1986 bringing the total core area of STR measuring 845.70 sq km declaring intention of the Govt. to form Similipal National Park. Similipal Forest Development Corporation (SFDC) was created to exploit the timber strictly as per Working Plan prescriptions with Departmental Officers in order to make good of the loss of habitat by working through contractors. The Corporation exploited timber and NTFP from 1979 to 1988 after which complete moratorium on tree felling was imposed. There are 65 villages located inside the sanctuary and about 1200 villages on the periphery with sizable population who dependant on the forest to meet their day to day requirement and supplementing their income through collection of NTFP items. As per Man and Biosphere (MAB) programme of UNESCO the Similipal Biosphere Reserve covering an area of 5569 sq km was created.

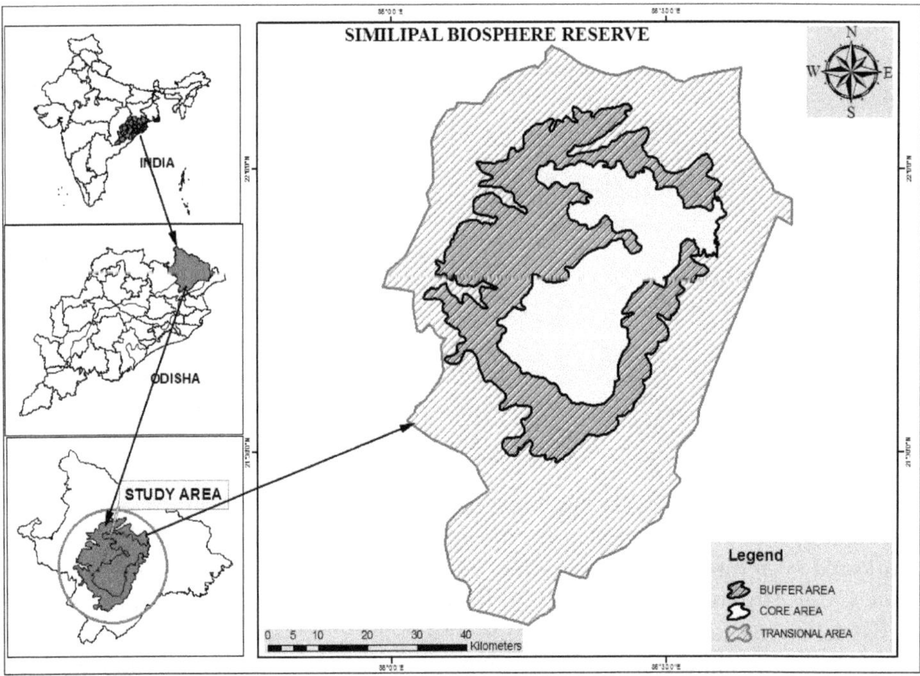

Figure.2.1. Map showing the location and zones of Similipal biosphere reserve

The Similipal has been declared country's eighth biosphere reserve and only in the state is not only one of the unique biodiversity rich areas of Odisha but also of the whole country. It is located in the northern boundary of the state as a north south elongated lenticular area in the district of Mayurbhanj between 21^0 16′ to 22^0 08′ N latitude and 86^0 03′ to 86^0 37′ E longitude (Fig.2.1). It is not only encompassing the Mayurbhanj district in its entirety but also the district of Balasore in the south-east, Keonjhar in the south-west, Midnapore district of West Bengal in the north-east and Singhbhum district of Jharkhand in the north-west, respectively. The biosphere reserve also surrounded by small townships of Udala and Thakurmunda in the south, Bangirposi and Bisoi in the north, Baripada in the east and Jashipur and Karanjia in the west (Fig.2.1). The reserve spreads over an area of 5569 sq km and includes within it a proposed national park, a wildlife sanctuary and greater parts of tiger and elephant reserves. The whole area of the reserve is divided into three different zones with a core area of 1194.75 sq km, buffer area of 1335.86 sq km and transitional area of 3038.39 sq km.

According to the biogeography of Rodgers and Panwar (1988) the three hierarchical levels of planning units under which Similipal has been classified are Deccan Peninsula, Chhotanagpur Province and Mahanadian region. However, it represents features of four biotic provinces for which Odisha is the junction. These provinces are Eastern Plateau,Chhotanagpur, Lower Gangetic Plain and Coastline. It is also considered as Himalaya of Odisha because of its enormous influence of the climate of the state and neighbouring states. It acts as a barrier to south-western monsoon causing heavy rain and diverts the moisture laden wind in south-western direction to cause rainfall in western Odisha. The western part of Odisha does not suffer from any drought because of the hill ranges of Similipal.

Topography

The topography of Similipal is undulating varying from small hillocks to high mountain peaks. The hills and hillocks rise abruptly from the coastal plains from southern (Udala) and eastern (Baripada) sides and rise steeply to height of 700 to 800 m M.S.L., extend towards north (Jashipur) and north-western side and merge with the chhotanagpur plains. On the western side there is the existence of large number of isolated hills. In the basin of the hills lie numerous valleys supporting habitations, meadows, etc. The two highest peaks of the reserve, i.e., Khairiburu (1168 m) and Meghasani (1165 m) stand close to each other and looks like two towers of the natural grandeur. The elevation of the central region of the plateau near Dhudurchampa is approximately 1100m. The elevation of few other points of the reserve are Jenabil (865m), Upper Barakamda (824m), Bhanjabasa (706m), Chahala (770m), Nawna (730m), Bakua (926m), Joranda (681m), Nigirdha (830m), Solamundi (850m), Ganapati (1114m), Berahuri (1092m), Amjhori (1082m) and Bisaldanga (1080m). The well wooded rolling plateaeu dotted with numerous hills and hillocks and having a network of perennial streams, rivulets and rivers are covered with dense vegetation. The two mighty waterfalls like Barehipani (217m) and Joranda (181m) which increase the beauty of the reserve is also the tourist attraction points of SBR. Besides this many rivers of north odisha are arise from them.

Water Resource of Similipal

A number of perennial rivers and their tributaries have originated from the Similipal hills. Budhabalang or the Balang, the largest river of Mayurbhanj and one of the largest river systems of Odisha rises from Similipal hills in double falls at Barehipani and flows into the Bay of Bengal. It has major tributaries like Palpala, Chipat, Gangahar and Nalua all are arise from Similipal. Sono, another important river of Odisha also arise from Similipal. Its major tributaries are the Kalanala, the Deo and the Sanjo all of are arise from Similipal. River Khairi takes its origin from the Similipal hills and finally meets the Subarnarekha river. From the southern slopes of the Meghasani peak rises another important river known as Salandi which flows into the bay of Bengal after joining Dhamra. The East Deo and West Deo are also originates from Similipal hills and meets Baitarani in Keonjhar district. Two other rivers namely Kharkei (Khaira) and Bhandan originate from the Similipal hills but they meet with each other at Jashipur. After their meeting at Jashipur, they jointly form the river Khairabhandan onwards. Besides these major rivers and rivulets Similipal is also the source of numerous hill streams. There is no lake in Similipal hills. Numerous lentic water bodies like ponds and tanks are mainly found in the plains. Thus Similipal is the source of both surface water and groundwater of the inhabitants of the district of Mayurbhanj and the three adjoining districts, i.e., Balasore, Bhadrak and Keonjhar.

Climate

The area enjoys a tropical climate with three distinct seasons, i.e., summer, winter and rainy. Summer extends from March to May, monsoon or rainy season from June to September and winter from November to February. Summer is moderately hot and is bearable as the temperature rarely goes above 38^{0}C. Thunderstorms are mostly common in the afternoons. The mean temperature of the reserve varies between 20^{0}C to 28^{0}C with a mean minimum range of 7^{0}C to 11^{0}C and a mean maximum range of 33.4^{0}C to 37.40^{0}C (Fig.2.2). Occasionally heavy rains occur during pre and post monsoon period due to depressions in Bay of Bengal. Most of the rainfall occurs during monsoon, i.e., during mid June to October accounting for 75-80% of annual rainfall, but the rainfall is not well distributed throughout the area. South Similipal receives higher rainfall than north Similipal and there is general decrease of rainfall from south-west to north-east. The atmosphere of the south-eastern part of Similipal remains humid due to the flow of sea wind. Meghasani and Khairburu hills obstruct free flow of moisture-laden wind giving more precipitation and creating different microclimate in this region. Starting of winter generally marked in November and becomes severe in December and extends up to February. During the peak of winter temperature comes down to 3^{0}C with frosts in valleys. Spring was very pleasant. Because of luxuriant vegetation cover and a network of perennial streams, Similipal is relatively moist throughout the year. Minimum humidity of Similipal ranged from 40 to 70 % at 06: 00 hrs and maximum ranged from 73 to 90 % at 18:00 hrs (Fig. 2.2).

Geology and Soil

Similipal was a part of Gondwana land in the Palaeozoic era. The main formation has three concentric bowls of impervious metamorphic rocks with their interfaces filled with pervious volcanic rocks made up of spilitic lava. These formations dip downwards the centre forming a basin structure. The typical bowl like formations of impervious and pervious rocks helps in holding of large amount of groundwater that feeds the river system and waterfalls of the reserve.

The metamorphic rocks are granitoid genesis, true genesis and micaschists with pegmatite. The geneissic rocks are much interspected by dykes of basic and intermediate rocks. The submetamorphic rocks are shale, haematitic rock laterites, limestone, calcareous deposits, quartzites, phyllites and micaceous schists. Haematitic rocks, laterite and shale occur in extensive formations in central and south Similipal. Outcrops of submetamorphic and quartzite haematite occur all over Similipal hills.

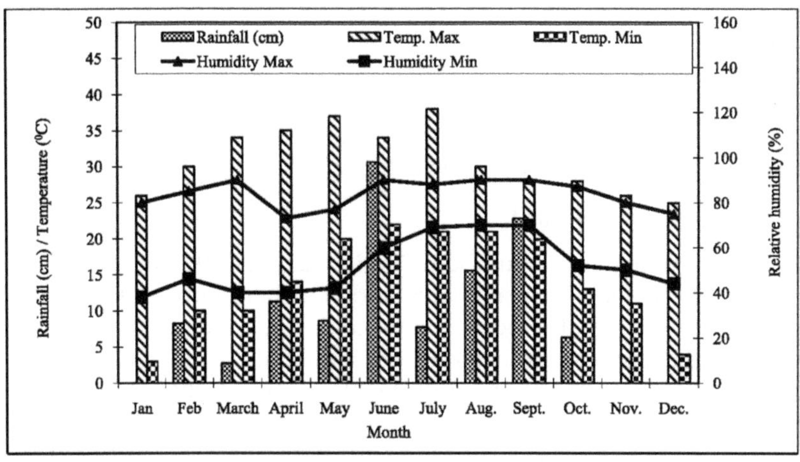

Fig. 2.2. Climatic feature of Similipal biosphere reserve.

The soil of Similipal is acidic in nature (pH range of 5.23 to 6.52 in most areas), red loam, derived from haematite rocks. Soils are also of clay and clayey loam type, formed due to weathering of shales. The outcrop of submetamorphic sandstone and quartzite hamatite's, on disintegration, produce reddish sandy soil. Presence of such type of soil in Similipal provides capability to support good tree growth. Places inside SBR where the depth of soil is less, it supports poor tree growth but good grass coverage for animals. Soil erosion has recently emerged as a problem, particularly around the valley. The organic carbon contents are rich in the soil, but the nitrogen and phosphorus are very poor.

Human Habitations

As a single compact forest the Similipal hills to the outsiders may be a dense forest full of wild animals and plants without any human habitation. But as soon as one enters into it a number of revenue villages along with vast stretch of agricultural land come to sight. The major communities inhabit inside the forest area are the Bathudi, Bhumija, Kolha, Santal, Ho, Munda, Gonda and Pauri Bhuyans. They live on agriculture, daily wage earning and collection of forest products such as fuel wood, sal leaves and seeds, kusum seeds, siali leaves and bark, honey, arrow root, lac, mahua flowers, mushrooms, etc. Two primitive tribal groups namely Khadias and Mankidias living inside SBR are still the foragers and food gatherers.

Road Connections

The centrally situated, heart shaped Similipal is connected by many a route from outside town. These routes have been constructed by the Forest Department for the supervision of forest operations. The departmental routes are: Baripada to nawna via Astiaghat, Pithabata, Lulung and Nigirdha, Jasipur to Nawna via Kaliani, Nanji Osara, Gurguria and Garh-Similipal, Jashipur to Nawna via Podagarh, Jamuani, Chahala and Barehipani, Bangirposi to Chahala via Talabandh, Thakurmunda to Nawna via Jenabil, Udala to Nawna via Bhanjabasa and Jenabil, Khunta to Jenabil via Dengam, Khunta to Jenabil via Dengam, Khunta to Jenabil via Ranigul and Makabadi, and Tangabila to Ramjhari via Sardha, Hatibadi and Dudhiani. Besides the roads constructed by the Forest Department several temporary ghat roads and foot paths are also visible throughout the hill ranges of Similipal. As such these foot paths are connecting only to some specific forest zones, food gathering people also use these foot paths for daily foraging activity.

Mineral Deposit of SBR

As mentioned in the Mauyurbhanj District Gazetter, 1967, in Similipal reserved forest area (near Barajori Pirhkata Pahar and Haticher) amphibolites belonging to the iron ore stage are found associated with banded quartzites. Quartzites belonging to Dhanjori stage are also available in the Similipal basin. Shales, Phyllites and Micaschists are found predominantly developed in Talbandh valley, Suliapat range, northern cliff face of Meghasani, Ganapat hill Deokund range and several other places. In the middle and northern part of the heart shaped Similipal Dalma volcanic including Epidiorites are noticed. Three district flows are found which vary in texture and appearance. Vasicular varities are common. In Nachuani hill, and Anjori hill, Gurguria, Similipal garh, , Kukurbaka, Chandanchaturi, Meghasani, Deokunda, etc. such typical exposures are seen. Gabbro-granophyre suite of rock and Gabbro-Anorthosites suite of rocks are also seen in several parts of Similipal Plateau. In the southern part of Similipal plateau and around Thakurmunda granites and granodiorites are seen. Newer dolerites are found developed abundantly in the Similipal basin. High level laterites are seen in Similipal plateau and other parts of Mayurbhanj district. As regards to the mineral deposits, the Similipal plateau is found rich in iron ore. This is mainly found in Gorumahisani, Badampahar and Suliapat. As the Similipal has already declared as a wildlife sanctuary, a tiger reserve and biosphere reserve and a proposed national park, mineral deposits have neither been surveyed further nor exploited.

Chapter-3

FOREST TYPES AND DYNAMICS OF SIMILIPAL

The total forest area of the state is 58,136 sq km which is 31.4% of the states geographical area and about 7.66% of country's forests. Of this, protected forests constitute about 26.7%, reserved forests 45.3% and unclassified forests 28% (FSI, 2006). Out of 15,524 sq km of protected forests, Similipal forest which has been declared as country's 8th biosphere reserve covers 5569 sq km. It is not only one of the unique forest ecosystems of Odisha, but also of the longest contiguous patch of Sal forests of the country. The tropical forest vegetation cover of India was first surveyed by Champion (1936) followed by Burtt Davy (1938), Hora (1950), Misra and Puri (1954) and Champion and Seth (1968). In Odisha during 1958, the Botanical Survey of India made survey of vegetation and flora of the district of Mayurbhanj, besides the floral survey of Haines and Mooney during 1924 and 1950, respectively. But the standard classification of the broad-leaved tropical forest types and sub-types of Similipal has been recorded by Champion and Seth (1968) as Northern Indian Tropical Moist Deciduous type. The keystone species of the area is Sal (*Shorea robusta*) ubiquitous all over the tract, but the type of Sal varies with the edaphic conditions and its associates also vary with the type and topography of the region. It varies from pure to almost pure crop in the hills and as well as in plains. Silvicultural system practised over the years of selective removal of non sal species, and excellent regeneration and establishment has resulted in having a pure crop of Sal. Abundance of grasses in valleys is another remarkable feature of the sal forests of Similipal. It is the result of regular annual fires which have probably kept the sal forests as stable pre-climax and in extreme cases have carried it further to the grassland in a process of retrogression. Because of the tropical climate, varied physiography and altitude, and soil properties, the vegetation of Similipal is much diversified rarely witnessed in a continuous compact forest patch. Another interesting feature is presence of distinct transition zone between different vegetation types. With respect to variation in floristic composition, physiognomy, life forms, climate, soil conditions and altitude, the main forest types in the Similipal hill ranges are as follows:

Northern Tropical Moist Deciduous Forests

This type extends over an area of about 1540 sq km below 70 metre height except in the moist and deep valleys towards the southern and the eastern aspects of Similipal. It occurs in portions in several compartments of the reserve namely Khairi, Batang West, Khadkei, etc (Table 3.1). The top storey in this type consists of predominantly deciduous species but the second storey has some evergreens. These forests are most extensive and occur all over Similipal often in continuation with the semi-evergreen type. By far the predominant species of this type of forest is sal which forms 50-90% of the crop. The height of sal depends on the site characteristics like gradients, soil depth, temperature, aspects and soil moisture. The canopy is open at places where heavy felling had been made in the past. Predominance of sal is due to aggressive and gregarious habit adaptability to varying soils and climates coupled with its resistance to grazing and burning, and power to regenerate itself with the later two factors. The top canopy of this type is represented by *Xylia xylocarpa* (Kongda), *Pterocarpus marsupium* (Piasal), *Careya arborea* (Kumbhi), *Bombax ceiba* (Simili), *Adina cordifolia* (Kuruma), *Dillenia pentagyna* (Rai), *Bridelia retusa* (Kasi), *Terminalia alata* (Asana), *Hymenodictyon excelsum* (Bhurkunda), *Michelia champaca* (Champa), *Anogeissus latifolia* (Dhaura), *Terminalia bellirica* (Bahada), *Lagerstroemia parviflora* (Sidha), *Grewa* sp., *Syzygium cumini* (Jamun), *Mitragyna parvifolia* (Mundi), *Croton roxburghii* (Putudi), *Buchanania lanzan* (Chara), etc. The middle storey is represented by *Kydia calycina* (Kapasia), *Cassia fistula* (Sonari), *Wendlandia* sp., *Alangium lamarckii* (Ainso), *Polyalthia* sp., etc. The under storey consists of mainly *Ardisia solanacea*, *Barleria strigosa*, *Flemingia* sp., *Leea* sp., *Indigofera cassioides* (Jheli), *Helicteres isora* (Modimodika), *Flemingia chapper*, *Strobilanthes* sp. and others. Climbers are heavy consisting of *Bauhinia vahlii* (Siali), *Dioscorea* sp., *Smilax* sp., *Millettia extensa* (Guadhuni), *Cissus* sp., *Ampelocissus* sp., etc. Herbaceous flora consists of *Curcuma* sp., *Zinziber* sp., species of Rubiaceae, Asteraceae and Acanthaceae, ferns and orchids. Among the grasses *Imperata arundinacea* is most common and also found are *Anthistiria gigantia* and *Symbopogon martini* and sabai grasses. *Thysnoclaena maxima* occurs near water courses, though it is not very common. In very moist places ferns and orchids are also met with. The common climbers are *Bahunia vahlii* (Siali), *Millettia extensa* (Guadhuni), *Smilax macrophylla* (Muturi), *Combretum decandrum* and *Dioscorea* sp. *Asparagus* is also found in this type of forest but not common. From the floristic composition given above, it will be found that *Bassia* sp., one of the common tree species of the district of Mayurbhanj is absent in this type of forest. *Indigofera pulchella* is found all over this type. It is more conspicuous in cleared sites and areas of overfelling. So also the case of *Imperata arundinacea* and *Flemingia chapper* is not very common. But the other indicator shrub that is the *Strobelanthus* sp., a species is believed to exist in moist conditions and its poor regeneration at comparatively higher places. Sabai is confied to a few limited places on the dry sides. *Cedrela toona* and *Trema orientalis* occur only on moist slopes and the latter gregariously in places where heavy felling had been made in the past. It is a fast growing and fast dying species and having a temporary phase in this type of forest. *Michelia champaca* (Champa) and *Mangifera indica* (Mango) occur in most valleys and are rather intrusions from the moist semi-evergreen forest of the valleys.

This forest remains almost leafless during the dry season. The trees are mostly with cylindrical boles possessing generally thick, fissured and fibrous bark which peels off as flakes. Such type of bark structure of trees of the moist deciduous forests of Similipal protects trees from fire. The crown of the dominant trees is generally spreading and rounded. This forest type thrives well in areas of Similipal where rainfall is below 2000 mm per year with a marked dry season extending up to 6 months. Such type of forests generally ascends from the plains up to an elevation of 70 metre. Where soil moisture conditions are poor, it spreads up to an altitude of 600m above mean sea level. In general it is very difficult for the area to understand the exact climatic, edaphic and physiographic requirements for the optimum growth of climax deciduous forest community because it thrives in varied localities when different permutations and combinations of various factors exist.

Ecologically this forest represents a transitional type from dry deciduous to semi-evergreen vegetation and are mostly confined to lower slopes of hills and occasionally extended up to 400 to 450m depending upon the favourable conditions. Total number of plant species documented in this type of forest cover of Similipal including tree species is 393, higher in comparison to the other forest types inhabiting inside Similipal. However, the mean tree density (557 plants/ha) and basal area (42.08 m^2/ha) observed in this type of forest cover is lower than that of the semi-evergreen forest types (Mohanty et al., 2005; Reddy et al., 2007). Anthropogenic pressure imposed in type of forest cover of Similipal is one of the major cause of disturbance was studied through the measurement of disturbance index which varies from 30 to 40% as reported by Mishra et al. (2006). Apart from such ecological information the soil physico-chemical characters and litter turn over rate of moist deciduous forest types of Similipal was extensively studied by Mishra et al. (2007). Pedological analysis in different moist deciduous forest sites of Similipal indicates that the soil texture in most of the areas varies from loam to clay loam and the colour from red to dusky red. Percentage of pore space and moisture content in such type of forest covers of Similipal mostly varies from 48 to 76 and 5 to 40 %, respectively (Mishra et. al., 2007). The soil nutrient parametres like organic carbon, nitrogen, phosphorus and potassium indicates the soil productivity status of any type of soil. Here the soil type of moist deciduous forest types of Similipal is productive as evidenced through the measurement of organic carbon, nitrate nitrogen and available phosphorus and potassium (Mishra et. al., 2007).

Northern Tropical Semi-Evergreen Forest

This type is characterized by a large number of evergreen tree species, heavy climbers and abundant epiphytes interspersed with deciduous tree species in the upper canopy. It is confined to smaller pockets mainly in the deep and damp valleys of the perennial streams and nalas towards the southern and eastern aspects of Similipal hills. The deciduous tree species found in the upper strata are leafless for a short period but the middle storey is almost entirely evergreen. This type extends over an area of about 80 sq km (Table 3.1). Although species found in the adjoining Moist Sal Forests (MSF) are also met with in semi-evergreen patches, the variability is fewer. Similarly fewer sal trees are also observed here. The ground storey is covered with evergreen shrubs. Due to the shade of the closed canopy, growth of grasses frequently observed in Moist Deciduous patches is absent. Due to evergreen undergrowth elements the impact of forest fire is considerably less compared to other zones. A few of the species of the top canopy and most of the under storey are evergreen and occur intimately mixed with each other. The upper storey comprises of species like *Michellia champaca* (Champa), *Artocarpas lacoocha* (Jeota), *Cedrela ciliate* (Toon), *Magniferra indica* (Amba), *Ailanthus excelsa* (Mahanim), *Bridelia retusa* (Kasi), *Litsaea nitida, Anthocephalus indica, Amoora rohituka, Syzyzium cumini* (Jamu), etc. However, in association to these species in the upper storey typical species found along the beds of the streams are *Salix tetrasperma* (Pani Begunia), *Trewia nudiflora* (Pani Gambhari), *Simplocos spicata, Diospyrus embryopteris* (Makar kendu), and at places *Terminalia arjuna* (Arjuno). A little higher up in damp areas the tree species met with are *Bombax ceiba, Alstonia scholaris, Ficus* spp. etc. and in still higher up areas are *Toona ciliata, Xylia xylocarpa* and *Sterospermum suaveolens*. The middle storey is represented by *Mesua ferrea* (Nageswar), *Polyalthia serasioides* (Champati), *Macaranga peltata, Litsea nitida, Glochidion* sp., *Biscophia javanica,* etc. Many evergreen shrubs and perennials such as *Leea crispal* and *Curcuma aromatica* (Palua) cover the ground flora. Grasses are rather rare because of the heavy shade of the closed canopy. The epiphytic flora is rich comprising several species of ferns, mosses, lycopods and orchids.

This type of forest covers of Similipal hills distributed in high altitude (400-1000m) areas receive more than 1200 mm mean annual rainfall and having mean tree density and basal area of about 665 plants/ha and 52.82 m^2/ha, respectively (Reddy et al., 2007) higher in comparison to the density (Plants/ha) and basal area (m^2/ha) of Northern Tropical Moist Deciduous Forest types of Similipal. However the total number of plant species including trees documented here less than Northern Tropical Moist Deciduous Forest types and higher than that of Dry Deciduous Hill Forests of Similipal (Reddy et al., 2007).

Dry Deciduous Hill Forests

It extends over an area of about 250 sq km mostly in the eastern and southern aspects of Similipal in steep and exposed slopes up to an elevation of 800 metres above mean sea level (Table 3.1). Climatologically it is remarkable though the regions of Similipal containing such type of forest covers get the maximum of rain, being first to interrupt the monsoon winds, supporting a dry mix type of forests with many of its species preferring towards xerophytic habitat. Such type of condition is clearly noticed due to poor soil moisture status. Here Sal forms about 25% of the crop and even more on suitable localities. It is generally found scattered along with other species of the type which are characteristic of the xerophytic conditions prevailing there. Many of the species found in this type of forest covers have long periods of leaflessness. Upper and lower canopies of such forests are composed of deciduous species. *Lannea coromondelica* (Moi), *Garuga pinnata* (Kathkusum), *Protium serratum* (Rimili), *Sterculia urens* (Dom Sal), *Sterculia villosa* (Udal), *Cochlospermum religiosum* (Kapasia), etc. The shrubs include *Cleistanthus collinus, Nyctanthes arbortristis* (Gangasiuli), *Gardenia latifolia* (Dambaru), *Randia* sp. *Ziziphus* spp., *Helecteres isora* (Modimodika), *Flacourtia ramontchi* (Bhaincho), etc. The climbers are not many and consist of *Acacia* sp., *Bauhina vahlli* and *Ventilago denticula*. Associated with sal in the upper strata tree species found are *Anogeissus latifolia* (Dhaura), *Terminalia alata* (Asana), *Terminalia chebula* (Harida), *Diospyrus melanoxylon* (Kendu), *Haldinia cordifolia* (Kuruma), *Cassia fistula* (Sonari), *Mallotus philippensis* (Sinduri), *Lagerstroemia perviflora* (Sidha), *Semicarpus anacardium* (Bhalia), *Buchanania lanzan* (Chara). Sal comes up very well in ferruginous loams and loamy clays. Epiphytes and ferns are not conspicuous. *Erythrina suberosa* has been found to grow on every dry ridge within this type. The area is subject to annual fires for which regeneration of woody perennials in this type of forest covers distributed in steep and exposed slopes of Similipal is not adequate.

On the basis of phytosociological observations of the vegetation of Dry Deciduous Hill Forests, Reddy et al. (2007) reported that number of species and their density are comparatively lower than the Northern Tropical Moist Deciduous and Northern Tropical Semi-evergreen Forest types of Similipal, but not drastically reduced. However, the basal area of tree species of Dry Deciduous Hill Forests is nearly equal to two times less than the basal area of Northern Tropical Moist deciduous and Northern Tropical Semi-evergreen Forest types of Similipal. It indicates that the climatic and edaphic conditions of Dry Deciduous Hill Forests of Similipal are not suitable for better growth and development of tree species.

High Level Sal Forests

This sub type is clearly recognizable in the Similipal hills and extends over an area of about 250 sq km above 800m altitude on the hill tops and plateaus (Table 3.1). The forest is characterized by almost pure stand of Sal (75%) with poor quality trees having stunted growth, epicarmic branches, knots occur mostly on laterite trap and crystalline rocks. This type of forests are found surrounding Dhudurchampa and Nedam in Balang-East block (Compartment no. 1,2,3,5,21 and 22) and in compartment no.7 of Sanjo block and West Deo-20, etc. These areas are subject to annual frost and fire. These abiotic factors make it difficult for sal to regenerate and establish. Due to annual fires and frost, such forests tend to become open and in extreme cases a steady retrogression to grassland takes place. Floristically the common associates of sal in this type of forest covers of Similipal are

Dillenia pentagyna (Rai), *Syzygium cerasoides* (Poijamu), *Cedrela toona*, *Careya arborea* (Kumbhi), *Pterocarpus marsupium* (Piasal), *Terminalia chebula* (Harida), *Terminalia alata* (Asana), *Wendlandia tinctoria* (Dhulda), etc. among trees, *Indigofera pulchela* (Gileri), *Colebrookia oppositifolia*, *Grewia* sp., etc. among shrubs and *Themeda* sp., *Imperata cylindrica*, *Chrysopogon* sp., etc. among the grasses. In open area subject to frost and fire *Strobilanthus* spp. gets replaced by grass mostly by *Imperata arundinacea* (Chana ghas), *Anthistria gigantia*, etc. In shallow soil *Phoenix acaulis* is found to grow.

Low Level Sal Forest

The characteristics feature of the sub-type which is fairly widespread and occupies most of the hill slopes up to an elevation of 400 metres and distributed in forest blocks of thickly populated localities like Pithabata, Udala, Dukura, Bisipur, Patbil, Tehasil, etc (Table 3.1). Sal forms the pure crop of this type of forest cover and is fairly wide spread and occupies most of the plain forest outside the Similipal reserve. But the dominancy of Sal is gradually dwindled due to heavy anthropogenic pressure imposed on these forest covers by the local inhabitants. Though regeneration of sal is adequate but its establishment is too difficult on account of plucking of leaves from the seedlings by the local inhabitants in the early stage of development, heavy grazing and annual fire. Edaphologically the area is suitable for the gregarious growth of Sal and is of yellow loam or red and rarely clay. The associates of sal are *Pterocapus marsupium* (Piasal), *Terminalia alata* (Asana), *Gmelina arborea* (Gombhari), *Madhuca indica* (Mohula), *Anogeissus latifolia* (Dhaura/Dhaw), etc. in the upper storey and *Ougeinia* sp., (Panjan), *Emblica officinalis* (Amla), *Cassia fistula* (Sunari) and *Buchnania lanzan* (Chara) in the middle storey. Shrubs found in this type of forest are *Carissa spinarum*, *Holarrhena antidysenterica*, etc. Bamboo is absent. Common climbers met with are *Bauhinia vahlii* (Siali), *Butea superba* (Noi Palasa), *Smilax zeylanica*, etc.

Grassland and Savannah

A very few selective patches of grasslands are found inside Similipal extending over an area of about 80 sq km above 900m elevation on hill tops and in the higher frosty valleys (Table 3.1). On hill tops, grasslands perhaps form the climax type of vegetation, but in the valleys it arises on account of biotic or natural factors. In such conditions sal and other frost sensitive species appears, but adverse climatic conditions in this area leads to their stunted growth. Frost is problematic for sal and other frost tender species to regenerate and establish. Factors like dying-back, frost lifting and blisters kill sal saplings up to 8 metres height. Retrogressive succession of vegetation cover and conversion of forest land to grassland in this area is facilitated by frost and annual fire. Among the tree species found in these areas are *Syzygium cerasoides* (Poijamu), *Symplocos racemosa* (Lodh), *Dillenia pentagyna* (Rai), etc. Among the grasses are *Themeda triandra*, *Themeda quadrivalvis*, *Themeda caudate*, *Themeda arundinacea*, *Themeda villosa*, *Cymbopogon flexuosus* (Dhanantri), *Imperata cylindrica* (Chana ghas), *Andropogon ascinodes*, *Apluda mutica*, *Heteropogon contortus*, *Cappilipedium assimile*, *Chrysopogon aciculatus*, *Chrysopogon verticillatus*, *Arundinella setosa*, *Artraxon* sp., *Bothriochloa* sp., *Iseilema laxum*, *Eragrostis* sp., *Cynodon dactylon*,

Sporobolus indicus, Digitaria sp. and a number of other herbaceous species. Scattered growth of *Symplocos* sp., *Syzygium* sp., is also found. Grassland and savannah of Similipal make it an ideal habitat for herbivores and elephants which abounds in the region. On the basis of habitat conditions in hills grass species found are: *Pollinidium angustifolium, Pogonatherum paneceum, Sehima nervosa,* etc., in moist areas the species found are: *Apluda mutica, Arundo donax, Eragrostis atrovirens, Phragmites karka, Sporobolus indicus* and *Sacciolepsis indica* and in other grasslands the species met with are: *Bothriochloa bladhii, Cymbopogon fresuosus, Cynodon dactylon, Heteropogon contortus, Imperata cylindrica* and *Themeda* sp.

As the grasslands are being gradually engulfed with sal trees, management strategy also aims at suitable habitat manipulation plans for their restoration in side Similipal. Management of grasslands inside Similipal is so essential for small herbivores, insects and other invertebrates, which contribute to the biodiversity of Similipal.

Plain Sal Forest

Sal (*Shorea robusta*) forms the pure crop of this type of forest and found mostly in very flat plains and in some slightly undulating areas of Similipal extending over an area of about 299 sq km. It occurs at an elevation of 25 to 75m above sea level in Baripada division and parts of Udala division, and at an elevation of 25 to 33m in Karanjia division (Table 3.1). Principal soil type of this forest is sandy loam or red-soil and rarely clay. On the basis of rainfall and humidity such type of forest cover is distributed in three distinct regions of Similipal viz. (i) the region having a heavy rain-fall and humid atmosphere due to proximity to sea as in Banahari-Muruda, Deoli, Udala and Kaptipada ranges which would correspond to type I (Northen Tropical Moist Semi-evergreen Forest), (ii) regions having a farley heavy rainfall and moist atmosphere as in Panchpir and Thakurmunda ranges corresponding to Northern Tropical Moist Deciduous forests and the region having far less rainfall than the other two and having definitely dry conditions as in Rairangpur, Bahalda and Roruan ranges corresponding to Dry Deciduous sal Forest. This subtype of Nortern Tropical Moist Deciduous forest is generally found in thickly populated forest blocks of Karanjia, Dudhiani, Badampahar and Rairangpur ranges. Though regeneration of Sal in these forest blocks is adequate but its establishment is so difficult due to various types of anthropogenic pressure imposed on these forest blocks by the local inhabitants in the form of indiscriminate tree felling, burning for charcoal, heavy grazing, shrub cutting, daily leaf and twig collection. The crop left is an even aged pure Sal coppice small pole crop with an even age up to 20 years all over the district expecting for those in Thakurmunda and Panchpir ranges where the crop is a little older consisting of big sized poles to middle aged trees due to extraction of older trees. Natural regeneration is generally very inadequate and in proximity of villages totally absent, the ground looking clean swept. The crop have been repeatedly cut back and have not been supplemented with natural regeneration consists of poor and unhealthy stems, two to three sprouting per stool and poor stocking. The heavily grazed areas, with a total absence of regeneration, are characterised by undergrowth of *Holarrhaena antidysenterica, Gardenia gummifera* and

Diospyros melanoxylon (Kendu). In such areas, if, forests once cut will never come back with the existing biotic factors of grazing, fire and indiscriminate cutting and leaf and twig collections, unless of course, these are eliminate and the crop is artificially aided. An extreme form of the retrogression will be found in Roruan Range where whole patches of sal forest have been reduced to mere bushes spreading on the ground. Such forests left to these areas will ultimately disappear. Common associates of Sal in this type of forest are *Terminalia alata*(Asana), *Diospyros melaoxylon* (Kendu), *Bassia latifolia, Buchnania lanzan* (Chara), *Anogeissus latifolia* (Dhaura), *Phyllanthus emblica* (Amla), *Prerocarpus marsupium* (Piasal) and *Cassia fistula* (Sunari). *Combretum decandrum* is to be seen in moist places only. *Carissa spinarum, Holarrhena antidys enterica* and *Croton* sp. are some of the commonest shrubs.

Dry Sal Forests

This type of forest is confined to parts of hill blocks outside the Similipal reserve namely Badampahar, Satkosia, Jari, Kanapat, Tunguru, Saranda, etc (Table 3.1). where though, the rainfall is high; the condition is not favourable for development of good quality Sal due to unfavorable edaphic and climatic factors. The forest blocks like Satakosia and Saranda receive heavier rainfall than others. But the rainfall within the forest blocks of Dry Sal Forests of Similipal is considerably lower than that of Northern Tropical Moist Deciduous Forests. It may, therefore, appears to be a transition from the Northern Tropical Moist Deciduous Forest type resulting from less rainfall on account of topographical position, dry soil conditions and may be due to past disturbances. It lacks in the presence of numerous perennial streams which helps considerably to keep the climate inside the forests cool and humid. Therefore, many of the deciduous tree associates of the dry type are commonly found along with Sal which is the predominant species forming about 40-60% of the vegetation. In this type of forest of Similipal, sal is found in good proportion but of poor quality (QIV) and height growth and unsoundness is common even in low age. The steep slopes and ridges of the hills where the drainage and moisture condition becomes acute sal gives room to associate species. Here the common associates of sal in the upper canopy are *Anogeissus latifolia* (Dhaura), *Terminalia alata* (Asana), *Diospyros melanoxylon* (Kendu), *Bassia latifolia, Adina cordifolia* (Kuruma), *Pterocarpus marsupium* (Piasal), *Lannea grandis, Madhuca latifolia* (Mahula), etc. The middle canopy is represented by *Cleistanthus collinus, Phyllanthus emblica* (Amla), *Gardenia latifolia, Gardenia gummifera, Wendlandia tinctoria* and *Ziziphus* sp., Common climbers found in this type of forest cover are *Bauhinia vahlii* (Siali), *Smilax* sp., (Muturi), etc. In Badampahar reserve forest the crop is even aged small poles due to *jhoom* cultivation and exploitation of mature trees in the past. In Notto and Satkosia reserve forests the crop is more uneven aged and the mature trees are very low in number due to their past exploitation. Fire is of annual occurrence and appears to be the main factor arresting progression of such type of forest to the moist Sal type. Due to heavy anthropogenic pressure in the present-day condition and jhoom cultivation practice in the remote past, the retrogression is caused instead of progression in this type of forest cover of Similipal.

Table 3.1: Summary of Forest Types of Similipal Biosphere Reserve (SBR)

Type No.	Name of the forest type	Area covered (Sq km)	Altitude (m)	Forest blocks/ Compartments/ Locations
2B/C3	Northern Tropical Semi Evergreen Forest	80	400-1000	Deep and damp valleys of perennial streams.
C3/C3	Northern Tropical Moist Deciduous Forest	1540	< 70	Sanjo-5, Sanjo-6, East Deo-8, Khadkei-1, Khadkei-2, Khadkei-3, Khairi-1 Khairi-2 , Balang West-1
C3/C2d	Plain Sal Forests	299	25 to 75	Deoli, Udala, Banahari-Muruda, Kptipada, etc.
5B/C1	Dry Sal Forests	234	–	Satkosia, Jari, Kanapat, Tunguru, Tanger, Saranda, Sarali, Badampahar, etc.
3C/C2e(i)	High level Sal Forests	250	800	West Deo-20, Balang East-1,2,3,5, 21 and 22, Sanjo-7, Pithabata, Udala, Betnoti and Deoli range.
3C/DS1	Grasslands and Savannahs	80	900	Tinadiha,Nawna, Jamuna, Upper Budhabalang basin, Jenabil, Dhudur Champa and Upper Barakamda (UBK).
5B/C1c	Dry Deciduous Hill Forests	250	< 400	Notto, on the steep eastern and southern faces of Similipal
3C/C2e(ii)	Low Level Sal Forests	–	400	Pithabata, Udala, Dukura, Bisipur, Patbil, Tehasil, etc.

Forest Dynamics

Tropical forests cover more than one fourth (27.6%) of world's land area, on which more than 500 million people depend on for their livelihood. Around 1980 there were 1937 million hectares of tropical forest on the earth, accounting for about

40 per cent of the tropics (FAO, 1988). It is said that these forests provide a habitation for about half the species of flora and fauna known. According to FAO (1990) forests are being destroyed at a rate of 17 million hectares annually. These forests have received much attention in recent years in comparison to other forests because of their species richness, high standing biomass and greater productivity (Bhat et al., 2000). These forests are also acted as the major carbon sink. However, the structure, function and composition of such forests undergo changes due to natural process or on account of human and live stock intervention. As a result, there is a lot of spatial and temporal variation in the reported values of species richness, composition and productivity. Continuous monitoring of forest stand on a long-term basis from time to time is useful to document the vegetation dynamics satisfactorily (Bhat et al., 2000). It describes the underlying physical and biological forces that shape and change a forest or the continuous state of change that alters the composition and structure of a forest. Forests change in structure and composition due to internal processes (competition between species for water, light and nutrients; differences in regenerative capacity of species, increasing tree age) and external processes (climate change, large scale disturbances, etc.). In a very general sense, undisturbed forests undergo a predictable succession in which hard wooded, long-lived, large seeded, slow growing species with specialized (mammal) seed dispersal replace soft-wooded, short-lived, small seeded, fast-growing species. During this process, forests increase in mean height, mean tree size and the number of leaf layers (i.e. the forest becomes darker).

Disturbance sets back succession or, in other words, it rejuvenates the forest. Disturbance can be characterised in three components: scale, intensity and frequency. The larger the scale, intensity and the frequency of disturbance, the larger is the impact on the existing forest vegetation and its inhabitants. Forest communities are more or less adapted to the disturbance regime that prevails at a certain site. In contrast, forests evolved in the absence of large-scale disturbances but with frequent small-scale single tree falls cause limited damage and are rich in hardwoods with heavy seeds that reproduces in late. A change in the disturbance away from the prevalent regime puts pressure on any forest ecosystem, as the species of plants and animals that are best adapted to the new conditions may be rare or even absent. Logging is a disturbance of high intensity (soil disturbance and compaction, gap creation), on a large scale and of moderate frequency (on a scale of decades). It will have a relatively low impact on forests that are regularly subjected to large disturbances, but a strong impact on forests, which are adjusted to low dynamics. In this way in tropical forests there is an assemblage between trees of different age classes and they forming different layers and regular conversion takes place from the lower age class to the higher age class to maintain a complex climax forest with inclusion of natural disturbances.

However, the human induced pressure in such type of forests create imbalance among the different age groups of tree species and disrupting the climax, which can be explained as the forest dynamics. Similipal biosphere reserve included under the Eastern Ghat region of India harbour a large variety of species and experiencing human and livestock induced disturbances. Data concerning stand structure, composition and dynamics on the long-term basis are scanty. Apart from forest working plans for the extraction of timber for commercial use and urban supply and few studies pertaining to volume increment most of the studies in relation to stand structure, phenological and demographic attributes of tree species and litter dynamics of the forest floor of Similipal was studies by Mohanty et al. (2005), Mishra et al. (2001, 2002, 2003, 2006 and 2008) and Reddy et al. (2007). On the light of the above-mentioned studies conducted on Similipal, the forest dynamics of Similipal is discussed below.

Structural Dynamics of Tropical Forests of Similipal

The forest covers of Similipal is one type of the tropical seasonal forests of Odisha can be found at the elevation range from 25 to 1165m above mean sea level nsl. The distinct dry season of at least 4 months is believed to be the limiting factor for the forest covers of Similipal. Here generally most of the forest covers (Northern Tropical Moist Deciduous Forest, Northern Tropical Semi Evergreen, Low Level Sal Forests, Plain Sal Forests) generally occupy the areas of deep well drained soils along the stream banks and in moist sites while the dry deciduous forest covers (High Level Sal Forests and Dry Sal Forests) are on rocky soil on small hill tops or upper slopes of shallow sandy soils. Depending on the mosaic pattern of topography different types of forest covers inhabiting in Similipal is localized in different forest blocks and different altitudes. The area covered by different forest covers is also different. Not only they are differed from each other in abiotic and topographic factors but also they differ from each other in terms of their structure and species composition. In this context the dynamics of forest covers of Similipal is broadly discussed under three different respects i.e. compositional dynamics (change in species composition), structural dynamics (change in density and basal area due to selective girth class felling bench marked by the State Forest Department from time to time and change in forest area with respect to change in time period) as well as change in forest area due to different natural and anthropogenic disturbances and ecological dynamics.

Ecological Dynamics

Generally on the basis of different ecological aspects the forest dynamics can be divided into three categories: evolution, succession and fluctuation. Evolution is the process of developing a new form or species in plants or animals and add up to the biotic community such as the invasion of *Lantena camera, Parthenium* and other weedy species towards the buffer and core areas of the reserve as evidenced through the floristic (Saxena and Brahmam, 1989; Rout, 2003) and phytosociological (Mishra et. al., 2008; Reddy et al., 2007) investigations from time to time. The succession is the process of progressive or retrogressive changes in plant community. Here either due to natural or anthropogenic disturbance conversion of one type of forest cover to another type is mentioned in the working plan of Shree Sripal Jee (1953-1973) for the reserved forest of Mayurbhanj district including Similipal and by Bose and Das (1973-1993) and Mishra (1973-1993) after inclusion of reserved forest blocks of Similipal in Baripada and Karanjia division, respectively. Community fluctuation is the cyclical changes in plant community such as seasonal variation in grassland community or phases in ground cover under different forest types of Similipal. In the heavily destructive areas, *Imperata* grass usually comes to occupy the site, and this grass community will be maintained as long as it still has annual forest fire in the area. In highly disturbed but moister site, *Eupatorium odoratum* is the common pioneer species intermixing with shrubs and small saplings. Elephant grass (*Neyraudia reynaudiana*) and *Thysanolaena maxima* are also found in many places. If there is no annual forest fire, it will progressively succeed to the tree and shrub stage within few years. The less disturbed sites are commonly occupied by pioneer tree species such as *Croton oblongifolius, Mallotus* sp., etc. may be found in dense stand from their old clumps. If the deciduous forests of the reserve are completely protected for a longer period of time the forest type will progressively succeeded to the Dry Evergreen Forest Type. Fluctuations in the moist deciduous forest covers of Similipal are markedly seen both in the tree canopies and on the forest floor. During the dry period, all trees shade their leaves and stop growth to prevent high water loss. The leaves turn from green to brown, yellow and then drop on the ground in November or at the beginning of the cool dry season. Grasses and herbs on the forest floor are dry out due to excessive dry condition of the atmosphere and lack of water in soil. Forest fire commonly starts from December and consume all dry leaves and dead grasses and herbs on the forest floor. The ground covers will grow as early as the beginning of the rainy season before the tree layers close the crown with the leaves. Most of the trees in this community have their flowers during the dry period and drop their fruits and seeds in the early rainy season. Seeds of the plants resistant to forest fire can germinate better under fire pressure. However, the seeds of the dominant tree species of Similipal, i.e., Sal can't germinate and its seedlings in such conditions are gradually diminished and replaced by other species. Herbs and grasses on the forest floor profuse their growth replacing the different varieties of seedlings of upper strata tree species. The species of high light demanders usually germinate and bud in very early rainy season before the trees in canopy layers have leaves. They can complete their life cycle within few weeks. While the more shade tolerant species germinate and begin their life-cycle late in the mid rainy season when the crown covers are fully mature. The seedlings will use at least 10 years to form the full grown saplings, 70 years to produce trees of 90 cm gbh, 105 years to produce 120 cm gbh, 125 years to produce 135cm gbh and 145 years to produce trees having 150 cm gbh as evidenced through the study of the statistics of growth and yield of tree species especially for *Shorea robusta* as briefly explained in working plan of Shree Sripal Jee prepared by the state forest department. So the forest structure is drastically changed due to the disappearance of the seedlings of upper strata tree species due to different types of anthropogenic pressure imposed on the forest covers of Similipal.

Change in Species Richness (Species Dynamics)

The forest cover of Similipal has its composition of the deciduous species in a good proportion, but in most of the localities and all of the forest types a species may become predominant such as sal (*Shorea robusta*). Forest type wise tree species richness was 100 for Semi Ever green Forest Type, 121 for Moist Deciduous Forest Type and 76 for Dry Deciduous Forest Type (Reddy et al., 2007). Taking all variety of plant forms (herbs, shrubs, trees and climbers) into account, the number of plant species ranges from 202 to 393 in dry deciduous forests to moist Deciduous Forests, indicating that the Moist Deciduous Forests are more diverse at spatial scale than their counterparts. However, when the species richness was concerned in terms of number of species per hectare the species richness of tree species varied from 24 (in Moist Deciduous Forest Types) to 69 (in Semi Evergreen Forest Types) species among different forest types of Similipal. Through an elaborative study on the long-term biodiversity assessment of Similipal Biosphere Reserve from 1998 to 2001, (Mohanty, 2001) reported that the number of tree species per hectare was 33 in Moist Deciduous forest covers of Similipal. After 6 years the number of tree species per hectare in such type of forest covers of Similipal was 24 as indicated from the study of pjytosociological observations on tree diversity of tropical forest of Similipal by Reddy et al. (2007). The drastic reduction in tree species richness within an interval of 6 years may due to habitat destruction through anthropogenic disturbances and also the conversion of Moist Deciduous Forest Types of Similipal to Dry Deciduous Forest Types.

Structural Dynamics

Tree mortality and growth are continuous processes in the community dynamics. Along with it there is also considerable change in number of trees and stems lost at different time periods either due to selective felling after the plants attaining certain girth class or the overmatured and damaged plants demarcated by the State Forest Department or due to illicit felling of trees. Loss of stems in particular may not reduce the number of species rather it reduces or changes basal area and biomass. Overexploitation by way of creaming out matured and prematured sound trees resulted invasion of bushy and herbaceous elements to the forest covers.

(i) Girth Class Systems and Felling Operations

The Similipal biosphere reserve rich with magnificent trees is for the first few years being exploited by contractors. Commercially the process has been a successful one. However, the process is being made to carry on the exploitation in terms of girth class in accordance with the working plans on a scientific and systematic basis, so far as local circumstances permit. The chronological events of system of exploitation of the Similipal forests under the lease period and in the period of execution of working plans from time to time are described below.

1. From the very beginning, i.e., from the year 1904-1905 there was no systematic exploitation of the forest covers of Similipal. But the systematic exploitation procedure was started after Mr. C. C. Hart, the Forest Officer draw a working plan report in which he laid down the principles to be followed in the exploitations of the State forests. However, he did not plan out the details of exploitation. He only prescribed a selection and improvement felling with an immediate object of removing 6′ (180cm) and above girth sal trees which were considered as matured or overmatured and would otherwise deteriorate in value if retained for a longer period. So at first a 10 year lease period was introduced by him started from 1906 to 1916 and the lease was entitled to all trees having 180cm girth and above

except some tree species. Tree species which were not included under the 1st decadal scheme are: *Schleichera oleosa, Terminalia chebula, Strychnus nuxvomica, Terminalia belerica, Bombax ceiba, Ficus benghalensis* and *Terminalia alata*. Apart from these tree species all edible fruit bearing trees viz. *Madhuca indica, Buchnania lanzan, Diospyros melanoxylon*, etc., all bamboo clumps and the trees along the bank of streams and nullahs were also not included.

2. The work of exploitation was started and continued under the so, called selection system till 1919 when "The Borooah Company Limited" was formed which took over all the assets and liabilities of the original lessee M/S B. Borooah and Company with the consent of lessor. On a mutual understanding East Similipal range was withdrawn from the lease. With this alteration in the terms and conditions of the lease, the pace and quantum of exploitation in north, west and south Similipal ranges naturally became excessive. In the plains forests of Banhari-Murunda, Deoli and possibly those of the present Udala ranges were worked for big trees through contractors. The system of exploitation was the selection system without proper control over the yield and method of work. Under this system forests were very much damaged and all big trees were combed out. But it appeared that there was no working plan to regulate the work. The fellings were unplanned and haphazard and therefore the forests were very much more depleted than before.

3. Up to 1939 the Borooah Company had been greatly exploited both north and west Similipal ranges as per schedule but the exploitation in south Similipal ranges was behind schedule due to its inaccessibility. To compensate the loss, the exploitable girth class of sal trees during this period was marked 150cm and 135cm gbh and more in north and west Similipal ranges, respectively. This lowering of the exploitable girth in comparison to the previous agreement, i.e., from 1906-1916 made a way for heavier exploitation and provided a scope to re-exploit the areas already worked over in the past. In lieu of the above concession, the timber company agreed to do improvement in feelings and climber cutting in areas where they were allowed to take up re-exploitation down to 135cm girth. Due to exploitation considerable opening of the canopy had been made which was apparent in subsequent years leading to development of profuse ground flora. There was no marking rule based on silvicultural considerations except the fixation of exploitable girth and excluding the trees on the banks of nullahs.

4. After the expiry of over 30 years of lease period of the contractors, i.e., from 1906 to 1946, in 1946 Dasgupta's plan was strictly executed under a selection-cum-improvement system of working. In Dasgupta's working plan not only the exploitable girth class of the sal trees were fixed but also the exploitable girth class for other tree species fixed namely *Michelia champaca* (Champa), *Pterocarpus marsupium* (Piasal), *Gmelina arborea* (Gambhari), *Xylia xylocarpa* (Bankhira), *Adina cordifolia* (Koim), *Terminalia alata* (Asana), *Bombax ceiba* (Simili), *Mitragyna parviflora* (Gudi Koim), *Syzygium cumini* (Jamu), *Anogeissus latifolia* (Dhaura) and *Holoptelia integrifolia* (Charla).

Exploitable girth for sal, Jamu, Dhaura and Charla was fixed at 165 cm at breast height, for champa 210cm, for piasal, simul, gambhari, godikoim and bankhira 180cm and for asana 195cm. Apart from these above mentioned species, the other tree species had fixed girth class for exploitation was 150 cm. Result's of the Dasgupta's plan explained that actually the trees of exploitable growth were too few and it was seen that there was about one Sal tree of 150cm and over girth at breast height per 1.2ha of forests. Thus cleanings were not done and the fellings were purely governed by selection system without any real improvement fellings as analysed through the sale lists of past years that about 60% of the annual yield was from trees of 90cm to under 150cm girth many of which were sound. This was being done in the name of improvement. As a result of this the conditions of the forests deteriorated further.

5. Dasgupta's working plan was in force for a short period from 1947 to 1952 till a new plan written by Shree Sripal Jee for the reserved forest of Mayurbhanj district came into operation from 1953-1973. This plan covered all the reserved forests in the district of Mayurbhanj including the whole Similipal and divided into 5 working circles namely selection working circle, improvement working circle, coppice working circle, bamboo working circle and protection working circle.

6. Selection working circle was formed to comprise all the valuable hill blocks of the division namely Similipal hill reserve, Noto, Satkosia and a part of Badampahar block. The system adopted was selection-cum-improvement. Improvement marking was prescribed for trees down to 5′ to 6′gbh (150-180cm gbh) in Similipal felling series and to 3′ to 4′ gbh (90-120 cm gbh) in case of Noto, Satkosia and Jashipur felling series. In actual practice, the marking rules have been revenue oriented, with very little help for effective improvement of the crop or for establishment of regeneration. In some places there has been excessive removal of the approach class trees whereas they are neglected in other parts. Climber cutting was not done and cultural operations were almost forgotten. Even the prescription for retention of one third of exploitable sal trees was not followed rigidly. Very little improvement of the crop could, therefore, be effected in practice although the prescriptions of the plan aimed at such improvement.

7. Improvement working circle was constituted including all the plain forests of Thakurmunda, Kendumundi and better quality forests of Karanjia range consisting of young to middle aged crop with a prescription of light improvement felling under a 30 year felling cycle. Though the objectives of the plan was to bring these forests under uniform system in due course, the marking rules provided exploitable girths for the different species to be marked along with dead, dying and diseased trees without providing any lower girth limit. In practice the marking rules were interpreted as nothing different from those for the selection-cum-improvement fellings prescribed in the selection working circle, thereby removal of all mature trees even without any reservation. The crop left in these ranges consists of even-aged

sal poles and trees of 30-40cm in diametre (92-126cm gbh). In the absence of definite prescription at the end of the plan, the principal objectives for the working circle has been defeated. Due to lack of protection from fire and grazing establishment, regeneration has also been affected.

8. The coppice working circle will include all the areas previously worked under coppice system. It also includes major part of Badampahar and other hill forests in Rairangpur Division. The Rairangpur forests were previously excluded from any systematic exploitation.

9. The bamboo working circle included a few small bamboo plantations in Baripada and Rairangpur Divisions. They were not under any systematic exploitation and proposed to work under a selection system.

10. The protection working circle included the areas where no fellings were prescribed. The areas identified for this working circle found in the North-West of Karanjia Division and in parts of Rairangpur Division. The areas selected are close to grazing and any felling by the people.

11. Exploitable girth class fixed for various tree species selected in Similipal during this working plan except Noto, Satkosia and Jashipur Forest block are: 150cm or more gbh for sal, 210 cm gbh for champa (*Michelia champaca*), koim (*Adina cordifolia*), piasal (*Pterocarpus marsupium*), gambhari (*Gmelina arborea*), toon and mango (*Mangifera indica*), 180 cm gbh for simul (*Bombax ceiba*), godikoim (*Mitragyna parviflora*), bankhira (*Xylia xylocarpa*), Chhachina and charla, 165 cm gbh for bhurkunda, 120 cm gbh for asan (*Terminalia alata*) and sisoo (*Dalbergia sissoo*) and 90 cm gbh for panjan(*Ougeinia oogeinensis*). However, the exploitable girth class fixed for different tree species during the same working plan period of Shree Sripaljee in Noto, Satkosia and Jashipur Forest block are: 105 cm gbh for sal (*Shorea robusta*), gombhari (*Gmelina arborea*), bankhira (*Xylia xylocarpa*), asan (*Terminaliu alata*), mahula (*Madhuca indica*) and jamu (*Syzygium cumini*), 120cm gbh for piasal (*Pterocarpus marsupium*), simul (*Bombax ceiba*) and koim (*Adina cordifolia*), 90 cm gbh for sisoo (*Dalbergia sissoo*) and 75 cm gbh for dhaura (*Anogeissus latifolia*).

12. After expiry of Jee's plan it was revised during 1973-74 and separate plan for Baripada and Karanjia Divisions were compiled by S. Bose and R. Mishra, respectively. The sample stock enumeration for first time was done and exploitable girth limit for different species was fixed.

13. For Baripada Division the primary consideration in deciding the exploitable size has been the stage to which most of the trees can grow without developing unsoundness and the market preferences of big size timbers that are prevailing at the moment. Accordingly the whole Division was divided into three forest ranges. In one range Kuliana, Kachudahan, Champagarh, Deokund, Taldiha and Meghasani are included whereas Dangadiha and Noto are forming two separate ranges. The exploitable girths have been fixed for different tree species in Kuliana, Kachudahan, Champagarh, Deokund, Taldiha and Meghasani range are: 150 cm for sal

(*Shorea robusta*) and sisoo (*Dalbergia sissoo*), 210 cm for piasal (*Pterocarpus marsupium*), 270 cm for champa (*Michelia champaca*) and 195 cm for asana (*Terminalia alata*) and kasi (*Bridelia retusa*). In dangadiha range the exploitable girth for different tree species are: 105 cm for panjan (*Ougeinia oogeinensis*), 180 cm for gambhari (*Gmelina arborea*), bankhira (*Xylia xylocarpa*), gudikoim (*Mitragyna parviflora*), toon, charla, mango (*Mangifera indica*), bhurkund, simul (*Bombax ceiba*), kadam (*Anthocephalus cadamba*), chachina, rimili (*Protium serratum*) and jamu (*Syzygium cumini*) and for all other tree species excluding the above-mentioned species was 150 cm gbh. In Noto range exploitable girth class fixed for different tree species in comparison to the other two ranges of Baripada Division are: 105 cm for sal (*Shorea robusta*), 120 cm far piasal (*Pterocarpus marsupium*), 90 cm for sisoo (*Dalbergia sissoo*), 120 cm for gambhari (*Gmelina arborea*), bankhira (*Xylia xylocarpa*), asana (*Terminalia alata*), mahula (*Madhuca indica*), jamu (*Syzygium cumini*), dhaura (*Anogeissus latifolia*), simul (*Bombax ceiba*) and koim (*Adina cordifolia*) and for all other species excluding the above all species in the Noto range is 90 cm.

14. For exploitation of tree species of different girth classes, Karanjia Division is divided into two ranges, i.e., within the Similipal hills reserve and the other one is the Satkosia and Jashipur range. Within the Similipal Hill reserve the exploitable girth class fixed for various tree species are: 150cm gbh for sal (*Shorea robusta*), 270cm gbh for champa (*Michelia champaca*), 105cm gbh for panjan (*Ougenia oogeinensis*), 165cm gbh for asana (*Terminalia alata*), 120cm gbh for sisoo (*Dalbergia sissoo*), 180cm gbh for koim (*Adina cordifolia*), piasal (*Pterocarpus marsupium*), gambhari (*Gmelina arborea*), toon, mango (*Mangifera indica*), simul (*Bombax ceiba*), gudikoim (*Mitragyna parviflora*), bankhira (*Xylia xylocarpa*), chachina, charla and bhurkund, and 150cm gbh for all other species except the above-mentioned species. However, exploitable girth for different tree species in Satkosia and Jashipur range are: 90 cm gbh for sal (*Shorea robusta*) and sisoo (*Dalbergia sissoo*), 120 cm gbh for piasal (*Pterocarpus marsupium*), koim (*Adina cordifolia*), simul (*Bombax ceiba*), gambhari (*Gmelina arborea*), bankhira (*Xylia xylocarpa*), asana (*Terminalia alata*), mahula (*Madhuca indica*), jamu (*Syzygium cumini*) and dhaura (*Anogeissus latifolia*) and for all other species except the above-mentioned species in the Satkosia and Jashipur range is 90 cm gbh.

Marking Rules for Tree Felling

As per the Jee's working plan the following rules are prescribed for marking of trees for felling and apply to all ranges of Similipal except for Noto, Satkosia and Jashipur ranges.

1. No regular marking were to be done in grassy savannahs, open forests bordering savannahs, and areas subjected to frost damage. Except that only dead and dying trees were marked.
2. All trees, where normal or defective, which have attained their respective exploitable girths were marked for felling except the followings.

(i) In case of sal trees normally for every two sound sal trees marked, one sound and healthy sal tree having one foot bigger in girth than the exploitable girth was kept as reserve.

(ii) Unmarketable species of exploitable girths, if they were not interfering with the growth of any principal species or with the growth of very promising stems of the useful species were retained.

(iii) Trees standing in blanks inside the forests with no established regeneration of any kind of species to take their place.

3. All dead, dying, defective and diseased trees of underexploitable girths down to 90 cm gbh and down to 60 cm gbh in case of Noto, Satkosia and Jashipur ranges were also be marked for felling except for sal, in which case only dead, dying, defective and diseased trees, which would deteriorate in value, if retained further, were marked for felling. Secondly, they can also be marked for felling if their removal is beneficial to the growth of established regeneration or older growth of the principal species that are already there. In no circumstances however, normal or healthy sal trees or nearly such trees of underexploitable girth were marked for felling.

4. All trees of the inferior species of 90 cm gbh and over which were interfering with the growth of established regeneration or older growth of the principal or useful species were marked for felling.

5. If after marking under the rules, stems of 90 cm gbh and above are left too congested, a further marking with a view to thin the crop were made so that the most promising stems of the principal and useful species are left favoured.

6. All marketable dead trees of all species and of all sizes were also marked for felling. In spite of this well-mentioned marking rules, revenue-oriented marking were also carried out. Cultural operations were not mentioned. Lack of market for cleaning produce constituted a fire hazard. Silvicultural considerations and principles were not included due to lack of technical knowledge of the staffs employed.

(ii) Extraction of Timber Through Felling Operations

Over the years, many practices like selection system, uniform system, coppice-with-reserves, which are based on the concept of harvested trees through natural regeneration, had to be replaced progressively by clear felling with artificial regeneration system. The reasons are many, but important of them being anticipated better productivity and operational convenience. However, such type of practices lead to depletion of natural forests as evidenced through the extraction of timber from natural forest covers without maintaining any scientific rules and regulations. There is continuous evidence of gradual degradation of the forest stock due to demands on various types of forest produce principally fire wood and illicit fellings which have assumed serious proportion in some areas. Such degradation is obviously the result of removal of forest produce far in excess of that which is prescribed in the working plans. As far as the timber extraction from the natural forest covers of

Similipal is concerned since its inception gradually extraction of timber used for various purposes is increased and the ecological conditions of the forest covers of Similipal is ruined leading towards habitat fragmentation and very slow rate of transformation of lower girth class individuals of important tree species of Similipal to their higher girth classes as evidenced through the enumeration results of trees in different diametre classes as mentioned in different working plans(Jee's working plan from 1952-53 to 1973-74 for Mayurbhanj district, Das and Bose working plan for Baripada division and Mishra's working plan for Karanjia division from 1973-74 to 1993-94, and Bakhla's working plan for Baripada and Karanjia division from 1993-94 to 2015-16) from time to time. The amount of timber extracted in different periods through selection system, uniform system, and coppice-with-reserves is described here. The timber operation phenomena was conducted in four different ranges of the forest covers of Similipal viz. East Similipal range (of old Sadar division), North Similipal range (of old Bamanghaty division), West Similipal range (of old Panchapir division) and South Similipal range (of old Udala division) included under Western Working Circles by 8 contractors and those in Eastern Working Circle by 4 contractors during the 30 years of lease period. The total quantity of timber exported out of the state during the lease period was 1,77,263.5 cu ft or 5,021.63 cu m. There were also about 1, 85, 000 cu ft or 5, 240.79 cu m of sawn and converted timber stacked at Jashipur and Monda (Manada) depots as well as in the forests. The maximum amount of timber stipulated in the lease for annual exploitation was 1,00, 000 cu ft or 2832.86 cu m. It can be seen from the above that there was hardly any silvicultural considerations is marking trees under the lease. But with the granting of the 30 years timber lease, scientific and systematic management principles were not followed strictly and during this lease period creaming out the best available stems of sal through ruthless exploitation of the whole of the extensive Similipal forest was made. This lease was only for one tree species, i.e., sal. During the 30 years of lease period a fixed quantity of converted Sal timber was exported annually and it was ranged from a minimum of 3,00,000 cu ft or 8498.58 cu m to a maximum of 5,00,000 cu ft or 14,164.3 cu m. No enumeration had previously been done and yield prescribed was probably more than what the forests could yield by way of annual incremental growth on a sustained basis.

The work of exploitation was started and continued under the system of exploitation till 1919 when "the Borooah Timber Company Limited" was formed which took over all the assets and liabilities of the original lessee m/sb. Borooah and company with the consent of the lessor. During the lease period maximum permissible conversion of timber yield per annum was raised from 5, 00,000 cu ft or 14,164.3 cu m to 8,00,000 cu ft or 22,662.9 cu m. By 1939 Similipal reserves are 2166.48 sq km or 216648 ha. Approximately 3/4th of the area or 1, 62,486 ha were under the lease from which a maximum of 8,oo,ooo cu ft or 22,662.9 cu m of sawn timber per year was extracted.

This accounts for a yield of 2 cu ft per acre or 0.142 cu m per hectare per annum was taken out. It was also to be considered that the yield of 2 cu ft per acre or 0.142 cu m per hectare per year for the vast area of 4, 01, 684 acres (1, 62,486 ha.) was taken out from a very much limited area. The fellings were not controlled and regulated by any systematic plan. The result of haphazard fellings all over the area and fellings of larger number of trees through lease system and felling of trees through marking rules operated by the forest department of leads towards reduction in higher girth class individuals of important tree species and degradation of forest covers of Similipal. In this context study of population structure of dominant tree species in Similipal conducted by Mishra et. al. (2008) stated that the distribution of species and individuals in different age classes showed a linear decline from juvenile to mature age classes. The distribution of basal area showed a progressive increase from juvenile to mature age classes signifying that smaller number of mature individuals accounted for a greater percentage of basal area. However, some of the important tree species of Similipal experiences gap-phase type of population structure in disturbed and moderately disturbed sites of the reserve reflecting the degradation of forest covers in Similipal.

(Iii) Area Under Felling Operations

Forest covers of Similipal contain a fairly good quality of miscellaneous timber yielding plants including the economically valuable tree species viz. *Shorea robusta, Pterocarpus marsupium, Dalbergia sissoo, Gmelina arborea*, etc. was logged by the local inhabitants for their own benefit and by the timber contractors for commercialization purpose before the formulation of working plans. During this period the system of exploitation was not systematic and scientific. However, the systematic and scientific method of exploitation was started after the formulation of first working plan by Mr. C. C. Hart in 1885, the then Forest Officer during his visit to the state to inspect and advise on forest matters. For the better management, forest covers are divided into different felling series and the felling series into coups. This working plan was followed by the state up to 1904 and during this period girth class of exploitable trees was fixed but the area for exploitation was not marked. So the plain and accessible parts have nearly all been denuded of mature sal except in the plains of south-west where there is a very few evidence/no evidence on felling was marked due to unaccessibility and remote situation of this area. As a result of which greater portion of the mature sal in the south-western portion of the reserved forests and also in the central group of hill forests. Based on the formation of felling series the forest covers of Similipal have been very ruthlessly exploited unsystematically so the crop condition has deteriorated considerably. The number of sound mature trees is few and most of them are hollow and defective. Also here the proportion of different girth classes are not normal there being scarcity of trees of top girth classes with a preponderance of younger trees. Change in place and area of felling through formation of coups from time to time in Similipal is evidence from working plans prepared by the State Forest Department for the extraction of timber for commercial use and urban supply is resulted towards loss of diversity of plants as well as also decrease in plant volume.

Change in Basal Area (Growth Dynamics)

With remarkable variation in species richness (24 to 69 tree species/ha) and tree densities (527 individuals/ha in Dry Deciduous Forest Types to 665 individuals/ ha in Semi Ever green Forest Types), the basal area (52.82 m^2/ha) was higher in the

Semi Ever green Forest Types in comparison to the all other forest types of Similipal as reported by Reddy et al. (2007) through the phytosociological observations on tree diversity of tropical forests of Similipal. The study conducted by Reddy et al. (2007) is sporadic and not on the basis of disturbance gradient. Tropical forest covers, which are maximally prone to different types of disturbances, must need the study of growth dynamics in relation to disturbance for better management and future stability. In contrast to the study conducted by Reddy et al. (2007), Mishra et al. (2005 and 2008) reported that the basal area cover of tree species of moist deciduous forest covers of Similipal in different disturbance gradients on the basis of measurement of disturbance index as the percentage of damaged individuals to the total number of normal and damaged individuals varies from 52.36 to 84.86 m^2/ha. But in an average basal area of tree species of the whole biosphere reserve was 66.68 m^2/ha (Mohanty et al., 2005). Compilation of data on growth and yield of tree species of Similipal in terms of basal area and the number of individuals in different girth classes conducted by various workers in different time periods provides information on the reduction in basal area and number of individuals in higher girth classes with progressive advancement in time. Reduction in basal area and number of individuals of tree species in higher girth classes may be either due to logging of trees of selective girth classes marked by the Forest Department through the principles mentioned for felling of tree species in different working plans from time to time and due to the severe problem created by the local inhabitants and forest timber smugglers to meet their economic benefits. As far as the plant volume is concerned in different girth classes, lower number of individuals of higher girth classes accounting more volume in comparison to the higher number of individuals of the lower girth class individuals. But with respect to change in time period or with advancement in time period, plant volume is gradually reduced as evidenced through comparing the number of individuals of important tree species of 70cm gbh and more. Compilation of data on the number of individuals of important tree species in higher girth classes, which indirectly states about the plant volume extracted from the forest covers of Similipal from 1954 to 1994 mentioned in different working plans provides information regarding the gradual reduction of higher girth class individuals leads to decrease in tree volume from time to time. Very low number of individuals of important tree species in higher girth classes (> 70cm gbh) as mentioned in the working plan of Baripada Division from 1973 to 1994 prepared by Bose and Das and in the working plan of Karanjia Division for the same period prepared by R. Mishra in comparison to the number of individuals of important tree species in >70cm gbh class as mentioned in the working plan of the whole Mayurbhanj District from 1953 to 1973 prepared by Shree Sripaljee.

Chapter-4

STRUCTURE, COMPOSITION AND PHYTOSOCIOLOGY OF PLANTS

Tropical forests are highly productive, structurally complex, genetically rich and renewable genetic resources (Roy et al., 2002). The tropical deforestation contributes to increase in atmospheric CO_2 and other gases affecting the climate and biodiversity. Though such type of forests occupy less than 7% of the land surface, there have the higher distinction of harbouring 50% of all plant and animal species (Mayers, 1992). The rate of forest loss due to deforestation as reported by Food and Agriculture Organistion [FAO, (2001)] is 15.2 million hectare per year (Data from 1990-2000). Assessment of the plant diversity of forest ecosystems is one of the fundamental goals of ecological research and is essential for providing information on ecosystem function and stability (World Conservation Monitoring Centre [WCMC], 1992; Tilman 2000; Townsend et. al., 2008). It has attracted attention of ecologists because of the growing awareness of its importance on the one hand and the massive depletion on the other (Singh, 2002; Lewis, 2009). Out of sixteen major forest types of India (Champion and Seth, 1968), tropical forests occupy 38 % of the total forest area in India (Dixit, 1997). However, in Odisha forest ecosystems cover about 37.34% of the State's geographical area and about 7.66% of country's forests. Large population of the state utilizes various components of the forests for both commercial and subsistence purposes. In the past few decades, heavy human pressure have reduced the forested area in the state resulting in degradation and fragmentation of historically contiguous landscapes posing threats to plant diversity (Murthy et al., 2007). So it is essential to conserve the plant diversity and has become a major concern for much of society and for many governments and government agencies at all levels (Tripathi and Singh, 2009).

The Man and Biosphere Programme launched by United Nations Educational Scientific and Cultural Organisation ([UNESCO], 1971 as cited in Parker, 1984) aims at conserving the floral wealth in protected areas established by the Govt. of India in different states. Similipal Biosphere Reserve (SBR), a northern tropical moist deciduous type of forest (Champion and Seth, 1968) situated in the Mayurbhanj district of Odisha has over the years, played important roles in maintaining the climate and livelihood of local communities (Srivastava and Singh, 1997).

The National forest policy in India tipulates 33% of the total geographical area to be under forest. Large area of fertile forest lands have been converted to other land uses to meet the demand of growing population. In addition opening of the close forests due to deforestation has resulted in increase in soil erosion, landslides, floods and loss of biodiversity and wildlife habitats. At the global level similar situation is reported from Brazil, Malyasia, Indonesia, Africa and Central American countries where loss of wildlife habitat ranges from 40-80% (Puri, 1995). The tropical dry forest of Coasta Rica (Heinrich and Hurka, 2004) floristically very rich and diverse compared to the dry forests of Puerto Rico (Hare et al., 1997). Compared to other tropical dry deciduous forests of Eastern Ghats of India (Krishnannkutty et al., 2006) which are under various degrees of anthropogenic pressures, the SBR occupies strong ecological position in terms of species number and diversity. SBR is generally believed to be floristically rich, containing many varieties of plant life forms and medicinal plants as well (Saxena and Brahmam, 1989). Carefully compiled and up-to-date information on diversity and distribution status of these plant resources is however lacking. Though human-induced pressure, mainly through illegal chainsaw logging and access to non-timber forest products (NTFPs) is on the rise, some scientific studies were carried out in this contest by various workers (Mishra et al., 2006, Mishra et al., 2008; Reddy et al., 2007) to assess the plant diversity status of this reserve. For the conservation status of the biosphere reserve to be known and to allow for the sustainable management, there should be need of proper documentation of diversity status of various plant life forms and their distribution pattern inside the reserve. Knowledge of floristic composition, structure and distribution of angiospermic plants of this biosphere reserve is critical in this direction.

Methods of Study of Phytosociology of Plants

To study the plant diversity status, 18 study sites were selected in East, West, North and South directions inside Similipal Biosphere Reserve (SBR).The vegetation analysis was conducted during 2005-2008 for all the six layers of the forest, i.e., trees, climbers, shrubs, herbs saplings and seedlings. The species were identified with flora guides (Saxena and Brahmam, 1994-1996; Haines, 1921-25). The tree layer was analyzed by sampling 20 quadrats of 10 m x 10 m size at each site. The size and number of samples were determined using the method of Mueller-Dombois and Ellenberg (1974). The abundance, density and frequency were calculated for the species. Importance Value Index (IVI) was determined as the sum of the relative frequency, relative density and relative dominance for tree layer only. Raunkiaer's frequency class (1934) analysis was used to assess the rarity or commonness of the tree species (Hewit and Kellman, 2002). In this classification percentage frequency of the species was classed as A, B, C, D and E; where A represents rare (0–20%), B is low frequency (20–40%), C is intermediate frequency (40–60%), D is moderately high frequency (60–80%) and E is high frequency or common (80–100%). With this classification, the expected distribution of the species is A>B>C≤ ≥D<E. The distribution pattern of different species was studied using the ratio of abundance to frequency (Whitford, 1949). Trees were ≥ 30cm cbh (circumference at breast height), saplings were 10-30 cm cbh and seedlings were <10cm cbh (Knight, 1975). The shrub and herb layers were analyzed by randomly placing 20 quadrats of 5m x 5m size and 1m x 1m size, respectively at each site during the post monsoon season. The diversity index at each site was computed by using Shannon-Wiener information function (Shannon-Wiener, 1963) and concentration of dominance by Simpson's index (Simpson, 1949), evenness and richness index following Pielou (1975) and Margalef (1958) (as cited in Tripathi and Singh, 2009), respectively. The presence of climbers on trees affects their growth and development. They have been noted to suppress natural regeneration and delay forest recovery (Babweteera et. al., 2001). The presence or absence of climbers on the trees was scored on a 5-point scale (Alder and Synnott, 1992) whereby 1, 2, 3, 4 and 5 represented trees that were: having bore climbers; trees over grown with climbers; climbers on the stem only; climbers in the crown only and climbers both on the stem and crown, respectively.

Floristic Compositions and Occurence

A total of 266 species belonging to 204 genera and 76 families were recorded from the study area, out of which 117 were tree species, 17 climber, 31 shrub and 101 herb species. Thus only approximately 24.72% of the estimated flora of Similipal (Saxena and Brahmam, 1989) was covered in the study (Table-4.1, 4.2, 4.3, 4.4, 4.5 and 4.6). A majority of the families were represented by only two or less species. The most common families were the Euphorbiaceae and Rubiaceae, each represented by 19 species; followed by Fabaceae (15 species), Mimosaceae = Acanthaceae (12 species each), Asteraceae (11 species), Cyperaceae= Moraceae= Caesalpinaceae = Combretaceae (9 species each), Malvaceae = Melastomataceae = Rutaceae = Poaceae (7 species each), etc. The average number of species per hectare was 74. The number of species per genus was 1.3 and that per family was 3.5. Species in various groups of plant life forms had a wide range of occurrence, ranging in frequency from 5-72% in herbs, 5-94% in shrubs, and 5- 100% in case of trees, climbers, saplings and seedlings (Table 4.1, 4.2, 4.3, 4.4, 4.5 and 4.6).

The species richness of a forest ecosystem depends on the number of species per unit area; the more species there are per unit area, the higher the species richness. A total of 266 species per 3.6 ha or 74 species per ha in the SBR is more or less similar compared to the number of species reported by several workers in other tropical forest covers of India (Parthasarathy, 1999, 80 to 85 species per ha in Kalakad Mundanthurai Tiger reserve; Parthasarathy and Karthikeyan, 1997, 57 species per ha in Mylodai-Courtallum reserve forest) and also 70 to 80 species per ha that have been observed in other studies in West African tropical high forests (Lawson, 1985; Vordzogbe et al., 2005). The species richness in neotropical forests showed a wide variation, ranging from 20 species per ha in Varzea forest of Rio Xingu, Brazil (Campbell et al., 1992) to 307 species per ha in the Amazonian Equator (Valenica et al., 1994).

In the old world tropics species, richness ranged from 26 species per ha in Kolli hills of India (Chittibabu and Parthasarathy, 2000) to 231 species/ha in Brunei Darussalam of South East Asia (Poulsen et al., 1996). In tropical rain forests, the range of species count per hectare is about 20 to a maximum of 223. The number of species in Similipal biosphere reserve was 74 per hectare and this number is at the lower side of the range given in tropical rain forests and neotropical forests. In the study of species richness of the western ghat, south India Sunderpandian and Swamy (2000) stated that pronounced dry season and relatively low annual precipitation factors may correlate with low species richness.

Raunkiaer's Classification of Vegetation Layers of SBR

In most of the plant life forms there were a high number of species that occurred only once. The distribution of the species into Raunkiaer's frequency classes showed that most of the species encountered were rare and very few species were encountered in intermediate and high or common frequency class (Table- 4.7). Except climbers all other groups of plant life forms does not follow the expected A>B>C ≥ ≤ D<E frequency distribution proposed by Raunkiaer (1934) as the number of species in frequency class D is higher than frequency class E.

Distribution Pattern

The distribution pattern of trees, shrubs, climbers, herbs, saplings and seedlings of the reserve is shown in Table-4.8. Odum (1971) stated that under natural conditions, a clumped distribution of plants is normal. A higher percentage of random and regular distribution reflects the greater magnitude of disturbance` such as grazing and lopping in natural forest stands. Most of the species of all the vegetational layers of the reserve showed generally clumped type of distribution in the present study. Regular distribution pattern is completely lacking in all the vegetation layers. Both in herb and seedling layers not a single species also showed random distribution pattern (Table-4.8).

Table 4.1: Families, species, density, basal area, frequency, distribution pattern and Importance Value Index (IVI) of trees in Similipal biosphere reserve.

Name of the family	Name of the plant species	Density (Plants/ ha)	Basal Area (m^2/ ha)	Frequency (%)	IVI	A /F
Rubiaceae	*Adina cordifolia* (Roxb.) Hook. f.ex. Brandis	11.94	1.44	77.78	6.54	0.04
Rutaceae	*Aegle marmelos* (L.) Corr.	4.72	0.353	44.44	2.81	0.05
Mimosaceae	*Albizia marginata* (Lam.) Merr.	3.33	0.37	22.22	1.80	0.14
Combretaceae	*Anogeissus latifolia* (Roxb. ex DC.) Wall ex. Guill	45.28	3.36	77.78	13.43	0.15
Bombacaceae	*Bombax ceiba* L.	10.83	1.53	55.56	5.64	0.07
Euphorbiaceae	*Bridelia retusa* (L.) Spreng.	4.72	0.45	38.89	2.73	0.06
Anacaediaceae	*Buchanania lanzan* Spreng.	19.44	0.81	66.67	6.16	0.09
Lecythidaceae	*Careya arborea* Roxb.	6.67	0.36	44.44	3.07	0.07
Flacourtiaceae	*Casearia graveolens* Dalz.	5.28	0.12	55.56	2.98	0.03
Caesalpiniaceae	*Cassia fistula* L.	6.67	0.12	44.44	3.06	0.07
Euphorbiaceae	*Cleistanthus collinus* (Roxb.) Benth. ex Hook.f.	3.33	0.06	11.11	0.94	0.54
Euphorbiaceae	*Croton roxburghii* Balak	6.11	0.21	22.22	1.92	0.25
Mimosaceae	*Dalbergia latifolia* Roxb.	2.78	0.21	22.22	1.41	0.11
Fabaceae	*Desmodium oojeinesis* (Roxb.) Ohashi	5.00	0.58	27.78	2.53	0.13
Dilleniaceae	*Dillenia pentagyna* Roxb.	29.17	2.53	77.78	10.24	0.10

Source: Mishra et al. (2012)

Name of the family	Name of the plant species	Density (Plants/ ha)	Basal Area (m^2/ ha)	Frequency (%)	IVI	A /F
Ebenaceae	*Diospyros embryopteris* Pers.	2.78	0.17	22.22	1.44	0.11
Ebenaceae	*Diospyros melanoxylon* Roxb.	8.61	0.66	44.44	3.73	0.09
Ebenaceae	*Diospyros montana* Roxb.	2.78	0.07	22.22	1.34	0.11
Burseraceae	*Garuga pinata* Roxb.	2.78	0.16	27.78	1.65	0.07
Simaroubaceae	*Gmelina arborea* Roxb.	5.28	0.44	50.00	3.21	0.04
Apocynaceae	*Holarrhena antidysenterica* Wall.ex A.DC.	2.78	0.07	22.22	1.31	0.11
Malvaceae	*Kydia calycina* Roxb.	5.56	0.22	38.89	2.51	0.07
Lythraceae	*Lagerstroemia parviflora* Roxb.	6.39	0.39	33.33	2.65	0.12
Anacardiaceae	*Lannea corromandelica* (Houtt.) Merr.	4.72	0.84	27.78	2.85	0.12
Sapotaceae	*Madhuca latifolia* Gmel.	12.78	1.074	44.44	4.84	0.13
Anacardiaceae	*Mangifera indica* L.	3.61	1.08	38.89	3.48	0.05
Magnoliaceae	*Michelia champaca* L.	7.78	0.98	11.11	2.78	1.26
Oleaceae	*Nyctanthes arbortristis* L.	3.89	0.15	27.78	1.78	0.10
Ochnaceae	*Ochna obtusata* DC.	5.83	0.35	16.67	1.87	0.42
Fabaceae	*Pterocarpus marsupium* Roxb.	12.22	1.24	55.56	5.43	0.08
Euphorbiaceae	*Phyllanthus emblica* L.	4.17	0.2	27.78	1.88	0.11
Burseraceae	*Protium serratum* (Wall. ex Colebr.) Engl.	32.22	1.99	83.33	10.08	0.09

Name of the family	Name of the plant species	Density (Plants/ ha)	Basal Area (m^2/ ha)	Frequency (%)	IVI	A /F
Mimosaceae	*Samanea saman* (Jacq.) Merr.	2.78	0.13	3.13	1.40	0.80
Sapindaceae	*Schleichera oleosa* (Lour.) Oken	13.33	1.48	33.33	5.05	0.24
Euphorbiaceae	*Securinega virosa* (Roxb. ex Willd.) Baill	11.11	1.02	33.33	4.13	0.20
Dipterocarpaceae	*Shorea robusta* Gaertn.f.	284.17	27.73	100.00	77.67	0.57
Myrtaceae	*Syzygium cumini* (L.) Skeels	23.06	2.03	83.33	10.19	0.07
Myrtaceae	*Syzygium cerasoides* (Roxb.)Chatt. & Kanjlal	18.06	1.27	66.67	6.64	0.08
Combretaceae	*Terminalia alata* Heyne ex Roth.	50.28	4.37	94.44	16.13	0.11
Combretaceae	*Terminalia bellirica* (Gaertn.) Roxb.	6.67	0.44	44.44	3.49	0.07
Combretaceae	*Terminalia chebula* Retz.	5.56	0.78	61.11	4.29	0.03
Verbenaceae	*Vitex leucoxylon* (L.f.)	5.56	0.21	38.89	2.50	0.07
Rubiaceae	*Wendlandia tinctoria* (Roxb.) DC.	3.61	0.21	27.78	1.82	0.09
Mimosaceae	*Xylia xylocarpa* (Roxb.) Taub.	2.78	0.2	16.67	1.27	0.20
Rhamnaceae	*Ziziphus mauritiana* Lam.	0.28	0.04	5.56	0.31	0.18
Rhamnaceae	*Ziziphus rugosa* Lam.	2.78	0.05	27.78	1.48	0.07
Total		793.67	71.043	-	299.75	-

Table 4.2: Families, species, density, frequency, abundance and distribution pattern of shrub layer in Similipal biosphere reserve.

Name of the family	Name of the plant species	Density (Individuals/ha)	Frequency (%)	Abundance	A/F
Euphorbiaceae	*Antidesma ghaesembila* Gaertn.	97.78	94.44	5.18	0.05
Myrsinaceae	*Ardisia solanacea* Roxb.	45.56	55.56	4.10	0.07
Violaceae	*Bixa orellana* L.	17.78	22.22	4.00	0.18
Rubiaceae	*Catunaregam spinosa* (Thunb.) Tirveng.	12.22	16.67	3.67	0.22
Meliaceae	*Cipadessa baccifera* (Roxb.) Miq.	3.33	5.56	3.00	0.54
Rutaceae	*Citrus medica* L.	48.89	44.44	5.50	0.12
Rutaceae	*Clausena excavata* Burm. f.	3.33	5.56	3.00	0.54
Verbenaceae	*Clerodendrum serratum* (L.) Moon	32.22	22.22	7.25	0.33
Euphorbiaceae	*Croton caudatus* Geisel.	20.00	27.78	3.60	0.13
Rubiaceae	*Gardenia resinifera* Roth	17.78	27.78	3.20	0.12
Euphorbiaceae	*Glochidion* sp.	15.56	22.22	3.50	0.16
Lamiaceae	*Gomphostemma parviflorum* Wall. ex Benth.	21.11	27.78	3.80	0.14
Tiliaceae	*Grewia hirsuta* Vahl.	22.22	27.78	4.00	0.14

Source: Mishra et al. (2012)

Name of the family	Name of the plant species	Density (Individuals/ha)	Frequency (%)	Abundance	A/F
Sterculiaceae	*Helicteres isora* L.	17.78	22.22	4.00	0.18
Euphorbiaceae	*Homonoia riparia* Lour.	22.22	44.44	2.50	0.06
Hypericaceae	*Hypericum gaitii* Haines	54.44	22.22	12.25	0.55
Rubiaceae	*Hyptianthera sticta* (Wild.) Wight & Arn.	144.44	66.67	10.83	0.16
Fabaceae	*Indigofera cassioides* Rottel ex DC.	125.56	50.00	12.56	0.25
Oleaceae	*Jasminum arborescens* Roxb.	10.00	27.78	1.80	0.06
Verbenaceae	*Lantana camara* L.	287.78	61.11	23.55	0.39
Vitaceae	*Leea asiatica* (L.) Ridsdale	15.56	22.22	3.50	0.16
Vitaceae	*Leea indica* (Burm. f.) Merr.	34.44	16.67	10.33	0.62
Melastomataceae	*Melastoma malabathricum* L.	122.22	50.00	12.22	0.24
Rubiaceae	*Pavetta tomentosa* Roxb. Ex Sm.	22.22	27.78	4.00	0.14
Lamiaceae	*Pogostemon benghalensis* (Burm. f.) Kuntze	341.11	55.56	30.70	0.55
Fabaceae	*Sesbania bispinosa* (Jacq.) W. F. Wight	18.89	22.22	4.25	0.19
Malvaceae	*Urena lobata* L.	46.67	38.89	6.00	0.15
Asteraceae	*Vernonia anthelmintica* (L.) Willd.	30.00	27.78	5.40	0.19
Lythraceae	*Woodfordia fruticosa* (L.) Kurz	188.89	77.78	12.14	0.16
Rubiaceae	*Gardenia gummifera* L. f.	31.11	22.22	7.00	0.32
Fabaceae	*Flemingia chappar* Buch. – Ham. ex Benth.	73.33	50.00	7.33	0.15
Total		1944.44	-	-	-

Table-4:3 Families, species, density, frequency, abundance and distribution pattern of herb layer in Similipal biosphere reserve.

Name of the family	Name of the plant species	Density (Individuals/ha)	Frequency (%)	Abundance	A/F
Malvaceae	*Abutilon indicum* (L.) Sweet	1944.44	16.67	23.33	1.40
Asteraceae	*Ageratum conyzoides* L.	1527.78	38.89	7.86	0.20
Acanthaceae	*Andrographis paniculata* (Bum.f.) Wall. ex. Nees	833.33	11.11	15.00	1.35
Commelinaceae	*Aneilema ovalifolium* (Wight) Hook.f.ex.	1472.22	11.11	26.50	2.39
Scrophulariaceae	*Bacopa monieri* (L.) Pennell.	2000.00	5.56	72.00	12.96
Capparaceae	*Cleome viscosa* L.	1277.78	11.11	23.00	2.07
Commelinaceae	*Commelina benghalensis* L.	1083.33	11.11	19.50	1.76
Commelinaceae	*Commelina palludosa* Bl.	583.33	5.56	21.00	3.78
Commelinaceae	*Commelina* sp.	611.11	11.11	11.00	0.99
Zingiberaceae	*Costus speciosus* (Koeing) Sm.	1555.56	5.56	56.00	10.08
Hypoxidaceae	*Curculigo orchoides* Gaertn.	6638.89	72.22	18.38	0.25
Zingiberaceae	*Curcuma amada* Roxb.	14138.89	55.56	50.90	0.92
Zingiberaceae	*Curcuma aromaticum* Salisb.	2111.11	27.78	15.20	0.55
Cyperaceae	*Cyperus rotundus* L.	7611.11	11.11	137.00	12.33
Cyperaceae	*Cyperus sp.*	1527.78	11.11	27.50	2.48
Fabaceae	*Desmodium trifolium* (L.) DC.	35888.89	27.78	258.40	9.30

Source: Mishra et al. (2012)

Name of the family	Name of the plant species	Density (Individuals/ha)	Frequency (%)	Abundance	A/F
Acanthaceae	*Dicliptera bleupleuroides* Mees.	861.11	11.11	15.50	1.40
Poaceae	*Eragrostis cilliata* (Roxb.) Nees	1805.56	44.44	8.13	0.18
Acanthaceae	*Eranthemum purpurascens* Nees	916.67	22.22	8.25	0.37
Convolvulaceae	*Evolvulus alsinoides* L.	21333.33	27.78	153.60	5.53
Convolvulaceae	*Evolvulus numularis* (L.) L.	14694.44	11.11	264.50	23.81
Cyperaceae	*Fimbristylis aestivalis* (Retz.) Vahl.	3138.89	11.11	56.50	5.09
Rubiaceae	*Knoxia sumatrensis* (Retz.) DC.	3861.11	11.11	69.50	6.26
Lobeliaceae	*Lobelia alsinoides* Lam.	888.89	11.11	16.00	1.44
Cyperaceae	*Mariscos* sp.	1500.00	11.11	27.00	2.43
Sterculiaceae	*Melotia curcurifolia* L.	777.78	5.56	28.00	5.04
Mimosaceae	*Mimosa pudica* L.	611.11	5.56	22.00	3.96
Orobanchaceae	*Orthosiphon rubicundus* (D.Don) Benn.	805.56	11.11	14.50	1.31
Urticaceae	*Pozoulzia pentandra* (Roxb.) Benn.	1111.11	16.67	13.33	0.80
Amaryllidaceae	*Pancratium trifolium* Roxb.	1722.22	22.22	15.50	0.70
Acanthaceae	*Phlogacanthus* sp.	611.11	5.56	22.00	3.96
Arecaceae	*Phoenix acaulis* Buch-Ham.ex Roxb.	888.89	22.22	8.00	0.36
Arecaceae	*Phoenix* sp.	666.67	11.11	12.00	1.08

Name of the family	Name of the plant species	Density (Individuals/ha)	Frequency (%)	Abundance	A/F
Euphorbiaceae	*Phyllanthus fraternus Webster*	3416.67	38.89	17.57	0.45
Acanthaceae	*Rungia pectinata* (L.) Nees ex DC.	1750.00	27.78	12.60	0.45
Amaranthaceae	*Celosia argentia* (L.)	583.33	5.56	21.00	3.78
Fabaceae	*Shuteria involucrata* (Wall.) Wt. & Arn.	583.33	27.78	4.20	0.15
Malvaceae	*Sida cordifolia* L.	1111.11	11.11	20.00	1.80
Rubiaceae	*Spermococe pusilla* Wall.	1527.78	5.56	55.00	9.90
Acanthaceae	*Strobilanthus auriculatus* Nees	888.89	22.22	8.00	0.36
Fabaceae	*Uraria picta* (Jacq.) Desv. Ex DC.	1138.89	16.67	13.67	0.82
Fabaceae	*Zornia diphylla* (L.)	1166.67	11.11	21.00	1.89
Rubiaceae	*Hediotys verticillata* (L.) Lam.	861.11	27.78	6.20	0.22
Orchidaceae	*Eulophia nuda* Lindi.	1027.78	33.33	6.17	0.19
Acanthaceae	*Barleria srigosa* (Wild.)	611.11	38.89	3.14	0.08
Cyperaceae	*Cyperus triceps* Endl.	861.11	27.78	6.20	0.22
Others	Others	14083.3	5.56 – 27.78	2 - 18	0.06 – 3.24
Total		1,69,500	-	-	-

Table 4.4: Families, species, density, frequency, abundance and distribution pattern of climber layer in Similipal biosphere reserve.

Name of the family	Name of the plant species	Density (Individuals/ha)	Frequency %	Abundance	A/F
Liliaceae	*Asparagus racemosus* Willd	9.44	38.89	4.86	0.12
Caesalpinaceae	*Bauhinia vahlii* wight. &Arn.	13.33	100.00	2.67	0.03
Fabaceae	*Butea superba* Roxb	3.33	33.33	2.00	0.06
Combretaceae	*Calycopteris floribunda* Lam	3.61	38.89	1.86	0.05
Combretaceae	*Combretum roxburghii* Spreng	5.28	55.56	1.90	0.03
Dioscoreaceae	*Dioscorea bulbifera* L.	8.06	38.89	4.14	0.11
Mimosaceae	*Entada rheedii* Spreng	1.39	16.67	1.67	0.10
Asclepiadaceae	*Hemidesmus indicus* (L.) R.Br.	6.39	44.44	2.88	0.06
Fabaceae	*Millettia extensa* (Benth.) Baker	8.33	27.78	6.00	0.22
Asclepiadaceae	*Pergularia daemia* (Forssk.) Chiov.	13.06	44.44	5.88	0.13
Liliaceae	*Smilax macrophylla* Roxb	8.89	38.89	4.57	0.12
Liliaceae	*Smilax prolifera* Wall ex Roxb.	0.56	5.56	2.00	0.36
Apocynaceae	*Aganosma caryophyllata* (Roxb.ex sims)G.Don	1.94	16.67	2.33	0.14
Euphorbiaceae	*Bridelia stipularis* Bl.	0.83	5.56	3.00	0.54
Lygodiaceae	*Lygodium flexicosum* (L.) Sw	1.39	11.11	2.50	0.23
Araceae	*Pothos scandens* L.	1.94	16.67	2.33	0.14
Oleaceae	*Jasminum flexile* Vahl	0.83	11.11	1.50	0.14
Total		88.6	-	-	-

Source: Mishra et al. (2012)

Table- 4.5: Families, species, density, frequency, abundance and distribution pattern of sapling layer in Similipal biosphere reserve.

Name of the family	Name of the plant species	Density (Individuals/ha)	Frequency (%)	Abun-dance	A/F
Rubiaceae	*Adina cordifolia* (Roxb.) Hook. f.ex. Brandis	18.89	44.44	2.13	0.05
Rutaceae	*Aegle marmelos* (L.) Corr.	6.67	11.11	3.00	0.27
Mimosaceae	*Albizia marginata* (Lam.) Merr.	12.22	11.11	5.50	0.50
Combretaceae	*Anogeissus latifolia* (Roxb. ex DC.) Wall ex. Guill	85.56	88.89	4.81	0.05
Barringtoniaceae	*Baringtonia acutangula* (L.) Gaertn.	11.11	16.67	3.33	0.20
Caesalpinaceae	*Bauhinia variegata* L.	8.89	16.67	2.67	0.16
Bombacaceae	*Bombax ceiba* L.	6.67	16.67	2.00	0.12
Anacaediaceae	*Buchanania lanzan* Spreng	58.89	55.56	5.30	0.10
Lecythidaceae	*Careya arborea* Roxb.	20.00	44.44	2.25	0.05
Flacourtiaceae	*Casearia graveolens* Dalz.	64.44	83.33	3.87	0.05
Caesalpinaceae	*Cassia fistula* L.	27.78	50.00	2.78	0.06
Cochlospermaceae	*Chochlospermum gossypium* DC.	7.78	5.56	7.00	1.26
Meliaceae	*Cipadessa baccifera* (Roxb.) Miq.	12.22	22.22	2.75	0.12
Euphorbiaceae	*Cleistanthus collinus* (Roxb.) Benth.-ex Hook. f.	18.89	16.67	5.67	0.34

Source: Mishra et al. (2012)

Name of the family	Name of the plant species	Density (Individuals/ha)	Frequency (%)	Abun-dance	A/F
Euphorbiaceae	*Croton roxburghii* Balak	14.44	27.78	2.60	0.09
Fabaceae	*Desmodium oojeinesis* (Roxb.) Ohashi	5.56	11.11	2.50	0.23
Dilleniaceae	*Dillenia pentagyna* Roxb.	50.00	61.11	4.09	0.07
Ebenaceae	*Diospyros malabarica* (Desr.) Kostel.	8.89	11.11	4.00	0.36
Ebenaceae	*Diospyros melanoxylon* Roxb.	21.11	33.33	3.17	0.10
Euphorbiaceae	*Glochidion lanceolarium*(Roxb.) Dalz.	8.89	11.11	4.00	0.36
Simaroubaceae	*Gmelina arborea* Roxb.	8.89	22.22	2.00	0.09
Sterculiaceae	*Helicteres isora* L.	10.00	22.22	2.25	0.10
Apocynaceae	*Holarrhena antidysentrica* Wall.ex A.DC.	7.78	16.67	2.33	0.14
Flacourtiaceae	*Homalium nepalens* Benth.	40.00	61.11	3.27	0.05
Malvaceae	*Kydia calycina* Roxb.	6.67	11.11	3.00	0.27
Anacardiaceae	*Nothopegia heyneana* (Hook. f.)	5.56	5.56	5.00	0.90
Lythraceae	*Lagerstroemia parviflora* Roxb.	17.78	27.78	3.20	0.12
Sapotaceae	*Madhuca latifolia* Gmel.	13.33	27.78	2.40	0.09
Annonaceae	*Miliusa velutina* (Dunal) Hook.f . & Thomas.	7.78	16.67	2.33	0.14
Rubiaceae	*Mitragyna parviflora* (Roxb.) Korth.	5.56	11.11	2.50	0.23

Name of the family	Name of the plant species	Density (Individuals/ha)	Frequency (%)	Abun-dance	A/F
Oleaceae	*Nyctanthes arber-tristis* L.	36.67	72.22	2.54	0.04
Euphorbiaceae	*Phyllanthus emblica* L.	33.33	61.11	2.73	0.04
Burseraceae	*Protium serratum* (Wall. ex Colebr.) Engl.	35.56	55.56	3.20	0.06
Sapindaceae	*Schleichera oleosa* (Lour.) Oken	45.56	66.67	3.42	0.05
Euphorbiaceae	*Securinega virosa* (Roxb. ex Willd.) Baill	38.89	50.00	3.89	0.08
Dipterocarpaceae	*Shorea robusta* Gaertn.f.	365.56	100.00	18.28	0.18
Sterculiaceae	*Sterculia urens* Roxb.	17.78	16.67	5.33	0.32
Bignoniaceae	*Sterospermum suaveolens* (Roxb.)DC.	13.33	16.67	4.00	0.24
Myrtaceae	*Syzygium cerasoides* (Roxb.)Chatt. & Kanjlal	18.89	33.33	2.83	0.09
Myrtaceae	*Syzgium cumini* (L.) Skeels	30.00	55.56	2.70	0.05
Combretaceae	*Terminalia alta* Heyne ex Roth.	74.44	88.89	4.19	0.05
Combretaceae	*Terminalia bellirica* (Gaertn.) Roxb.	18.89	27.78	3.40	0.12
Combretaceae	*Terminalia chebula* Retz.	11.11	33.33	1.67	0.05
Rubiaceae	*Wendlandia tinctoria* (Roxb.) DC.	20.00	33.33	3.00	0.09
Mimosaceae	*Xylia xylocarpa* (Roxb.) Taub.	15.56	27.78	2.80	0.10
Rhamnaceae	*Ziziphus rugosa* Lam.	7.78	22.22	1.75	0.08
Others		148.74	5.56-16.67	1.00-4.00	0.08-0.72
Total		1524.34	-	-	-

Table 4.6: Families, species, density, frequency, abundance and distribution pattern of seedling layer in Similipal biosphere reserve.

Name of the family	Name of the plant species	Density (Individuals/ ha)	Frequency (%)	Abundance	A/F
Dipterocarpaceae	*Shorea robusta* Gaertn.f.	27777.78	55.56	100.00	0.56
Euphorbiaceae	*Croton roxburghii* Balak	24250.00	145.50	33.33	4.37
Combretaceae	*Terminalia alata* Heyne ex Roth.	3416.67	9.46	72.22	0.13
Anacardiaceae	*Buchanania lanzan* Spreng.	4944.44	14.83	66.67	0.22
Ebenaceae	*Diospyros melanoxylon* Roxb.	5888.89	26.50	44.44	0.60
Euphorbiaceae	*Cleistanthus collinus* (Roxb.) Benth. Ex Hook.f.	4611.11	23.71	38.89	0.61
Sterculiaceae	*Sterculia urens* Roxb.	1861.11	22.33	16.67	1.34
Flacourtiaceae	*Homalium nepalens* Benth.	1611.11	8.29	38.89	0.21
Euphorbiaceae	*Phyllanthus emblica* L.	2138.89	12.83	33.33	0.39
Combretaceae	*Anogeissus latifolia* (Roxb. Ex DC.) Wall ex. Guill	1500.00	13.50	22.22	0.61
Rutaceae	*Aegle marmelos* (L.) Corr.	972.22	7.00	27.78	0.25
Mimosaceae	*Dalbergia sisoo* Roxb.	500.00	18.00	5.56	3.24
Sapindaceae	*Schleichera oleosa* (Lour.) Oken	805.56	4.14	38.89	0.11
Simaroubaceae	*Ailanthus* sp. Roxb.	805.56	29.00	5.56	5.22

Source: Mishra et al. (2012)

Name of the family	Name of the plant species	Density (Individuals/ ha)	Frequency (%)	Abundance	A/F
Apocynaceae	*Hollarhaena antidysentrica* Wall.ex A.DC.	1361.11	7.00	38.89	0.18
Oleaceae	*Nyctanthes arbortristis* L.	1527.78	18.33	16.67	1.10
Rhamnaceae	*Ziziphus rugosa* Lam.	916.67	5.50	33.33	0.17
Euphorbiaceae	*Bridelia retusa* (L.) Spreng.	1277.78	11.50	22.22	0.52
Rubiaceae	*Ixora* sp.	722.22	26.00	5.56	4.68
Rubiaceae	*Gardenia gummifera* .L.f.	361.11	4.33	16.67	0.26
Fabaceae	*Pterocarpus marsupium* Roxb.	916.67	5.50	33.33	0.17
Fabaceae	*Desmodium oojeinensis* (Roxb.) Ohashi	1194.44	10.75	22.22	0.48
Myrtaceae	*Syzygium cumini* (L.) Skeels	2472.22	7.42	66.67	0.11
Rubiaceae	*Wendlandia* sp.	2000.00	72.00	5.56	12.96
Sterculiaceae	*Helicteres isora*. L.	972.22	8.75	22.22	0.39
Mimosaceae	*Albizia odoratissima*. (L.f.) Benth.	1222.22	22.00	11.11	1.98
Bignoniaceae	*Sterospermum suaveolens* (Roxb.) DC.	638.89	7.67	16.67	0.46
Combretaceae	*Terminalia chebula* Retz.	861.11	5.17	33.33	0.16
Meliaceae	*Trichilia connaroides* (Wight & Arn.) Bentv.	638.89	23.00	5.56	4.14
Euphorbiaceae	*Securinega virosa* (Roxb. Ex Wild.) Baill	500.00	6.00	16.67	0.36

Name of the family	Name of the plant species	Density (Individuals/ ha)	Frequency (%)	Abundance	A/F
Flacourtiaceae	*Casearia graveolens* Dalz.	1750.00	6.30	55.56	0.11
Dilleniaceae	*Dillenia pentagyna* Roxb.	2194.44	11.29	38.89	0.29
Mimosaceae	*Xylia xylocarpa* (Roxb.) DC.	1583.33	19.00	16.67	1.14
Mimosaceae	*Dalbergia latifolia* Roxb.	805.56	29.00	5.56	5.22
Burseraceae	*Protium serratum* (Wall.ex Colebr.) Engl.	694.44	3.57	38.89	0.09
Mimosaceae	*Albizia marginata* (Lam.) Merr.	861.11	15.50	11.11	1.40
Rubiaceae	*Adina cordifolia* (Roxb.) Hook. F.ex. Brandis	361.11	6.50	11.11	0.59
Anacardiaceae	*Mangifera indica* L.	527.78	4.75	22.22	0.21
Sapotaceae	*Madhuca latifolia* Gmel.	583.33	10.50	11.11	0.95
Lecythidaceae	*Careya arborea* Roxb.	527.78	4.75	22.22	0.21
Euphorbiaceae	*Securinega virosa* (Roxb. Ex Wild.) Baill	583.33	21.00	5.56	3.78
Simaroubaceae	*Gmelina arborea* Roxb.	416.67	5.00	16.67	0.30
Mimosaceae	*Albizia procera* (Roxb.) Benth	388.89	14.00	5.56	2.52
Bombacaceae	Bombax ceiba L.	388.89	7.00	11.11	0.63
Euphorbiaceae	*Mallotus phillipensis* (Lam.) Muell.-Arg.	416.67	15.00	5.56	2.70
Lythraceae	*Lagerostroemia parviflora* Roxb.	333.33	12.00	5.56	2.16

Name of the family	Name of the plant species	Density (Individuals/ ha)	Frequency (%)	Abundance	A/F
Rubiaceae	*Wendlandia tinctoria* (Roxb.) DC.	416.67	15.00	5.56	2.70
Meliaceae	*Cipadessa baccifera*(Roxb.) Miq	333.33	12.00	5.56	2.16
Other		1444.44	1.00-9.00	5.56-11.11	0.18-1.62
Total		1,13,416.7	-	-	-

Table 4.7: Distribution of vegetation layers according to Raunkiaer's classification scheme (Values in parentheses indicate % of species).

Frequency Class	Code	Number of species in vegetation layers						
		Tree	Climber	Shrub	Herb	sapling	seedling	Remark
0-20	A	71 (61)	7 (41)	4 (13)	78 (77)	79 (73)	34 (59)	Rare
21-40	B	27(23)	6 (35)	16 (52)	20 (20)	13 (12)	18 (31)	Low
41-60	C	9 (08)	3 (18)	7 (22)	2 (02)	7 (06)	2 (03)	Intermediate frequency
61-80	D	6 (05)	0 (0)	3 (10)	1 (01)	5 (05)	3 (05)	Moderately high frequency
81-100	E	4 (03)	1 (06)	1 (03)	0 (0)	4 (04)	1 (02)	High frequency (common)

Table 4.8: Distribution pattern of vegetation layers of Similipal biosphere reserve.

Plant group	Number of species in distribution pattern categories			
	Regular	**Random**	**Contiguous**	**Total number of species**
Tree	0	08	109	117
Climber	0	03	14	17
Shrub	0	01	30	31
Herb	0	0	101	101
Sapling	0	06	102	108
Seedling	0	0	58	58

Distribution of Climbers

Out of 794 number of trees per hectare 110 number of trees per hectare affected by 40% bore climbers, 10 % were overgrown with climbers while 15% had climbers restricted to the main stem, 12% had climbers in the crown only and 23% had climbers both the stem and in the crown (Figure. 4.1).

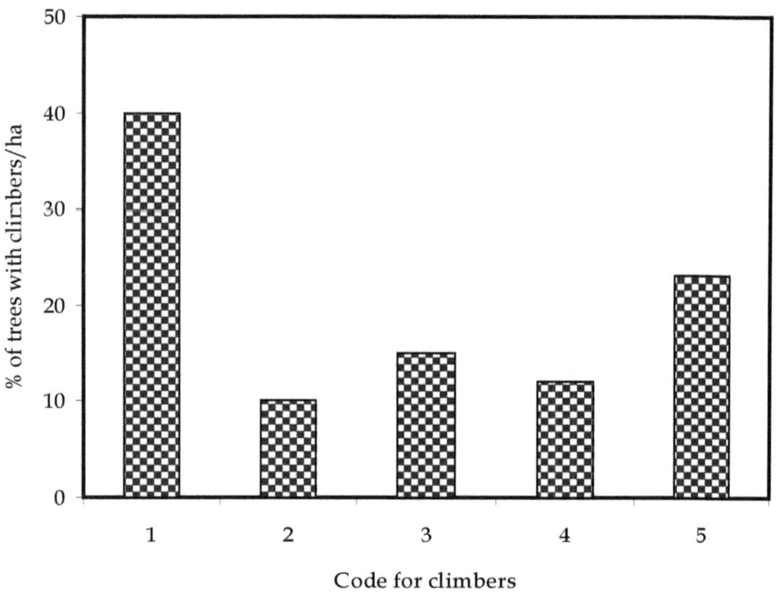

Figure 4.1: Percentage distribution of climbers in Similipal biosphere reserve.

The distribution of climbers on the trees of the reserve was considerably low, amounting nearly equal to 14%. This may be due to high canopy coverage, thereby allowing low light to reach the forest floor not triggering vigorous growth of climbers (Babweteera et al., 2001). Climbers impact on the vitality of trees negatively (Toledo-Aceves and Swaine, 2008) causing loss of foliage thereby reducing the surface area available for metabolic processes and reproductive potential as well as impeding or obstructing forest succession (Toledo-Aceves and Swaine, 2008). Notwithstanding their negative impacts, climbers form bridges between the forest canopies, thereby facilitating the movement of arboreal animals across the forest. They also protect weaker trees from strong winds (Schnitzer and Bongers, 2002).

Ecological Importance of Species

Importance Value Index (IVI) is the measurement of ecological amplitude of species (Ludwig and Reynolds, 1988) which is the ability of a species to establish over an array of habitats. However, there is no single perfect way of assessing the ecological amplitude of a species. The abundance of a species can be represented by several measures such as relative density, relative frequency and Importance Value Index (IVI). Though frequency and density values are suitable for herbs and shrubs (Airi et al., 2000), IVI is an important information for tree species. On the basis of IVI *Shorea robusta* was found as the dominant species in the SBR having IVI of 77.67 followed by *Terminalia alata* (16.13) and *Anogeissus latifolia* (13.43). *Wendlandia* sp. had IVI of 0.25 and is considered as the rare species of the reserve. All other tree species showed intermediate range of IVI (Table-4.1).

It has become common practice in quantitative descriptive studies to use IVI, which combines frequency, density and dominance into a single measure to analyze a plant community. Though vegetation can be described in terms of a number of parametres including frequency, density and cover, the use of any one of these quantitative parametres could lead to oversimplification or underestimation of the status of the species (Kigomo et. al., 1990, Oyun et. al., 2009). Except few tree species viz. *Shorea robusta, Terminalia alata* and *Anogeissus latifolia,* low ecological status of most of the tree species in the present investigation, as evidenced by the IVIs, may be attributed to lack of dominance by any one species, which suggests positive interactions among the tree species. In other words, resource spaces are shared to minimize negative species interactions and plants can obtain resources with relative ease (Tsingalia, 1990). The low IVIs may also imply that most of the tree species in this forest are rare (Pascal and Pellissier, 1996; Oyun et al., 2009), as confirmed by Raunkiaer's frequency distribution of the tree species. The rarity of species may be attributed to the occurrence of abundant sporadic species with low frequency in the stands (Oyun et. al., 2009). The high percentage (>70 %) of rare species observed in various vegetational layers of the reserve confirms the generally acclaimed notion that most of the species in an ecological community are rare, rather than common (Magurran and Henderson, 2003).

Stand Structure

Species wise density of individuals having $\geq$30cm girth of the reserve ranged from less than one plant per hectare to 284 plants/ha and the total density of the reserve was 794 plants/ha. Maximum density (per hectare of individuals of ≥30cm) was recorded for *Shorea robusta* (284) followed by *Terminalia alata* (50), *Anogeissus latifolia* (45), *Protium serratum* (32) and *Dillenia pentagyna* (29). Density was observed less than or equal to one for many species like *Antidesma acidum, Artocarpus lacuccha, Butea monosperma,, Casearia elliptica, Chionanthus intermedicus, Cochlospermum religiosum, Euonymus glaber*, etc. All other species showed intermediate range of density per hectare (Table-4.1). The densities of climbers in comparison to other vegetation layers of the reserve were too low. However the densities of herbs, seedlings and sapling layers were quite high in comparison to other vegetational layers. Unlike tree layer in herb, shrub and climber layers very few species showed lowest density. *Exacum bicolor* in herb layer, *Cipadesa baccifera* and *Clausena excavata* in shrub layer and, *Jasminum flexile* and *Bridelia stipularis* in climber layers were showed minimum value of density (Table-4.2, 4.3, 4.4, 4.5 and 4.6). Total basal area of trees of the reserve was 71.05 m^2/ha in which maximum was experienced by *Shorea robusta*. *Shorea robusta* contributed maximum of 39% to the basal area followed by *Terminalia alata* (6.15%), *Anogeissus latifolia* (4.73%) and *Dillenia pentagyna* (3.57%). The total contribution that resulted from this associated combination of *Shorea-Terminalia-Anogeissus-Dillenia* was 53.45%. A few families contributed most to the total basal area. These included the Dipterocarpaceae (39%), Combretaceae (13 %), Myrtaceae (5%), Rubiaceae (4.5 %) and Moraceae (4 %). As a whole the tree density and basal area of 794 plants/ha and 71.05m^2/ha, respectively is well within the reported range of various Indian tropical forests (Visalakshi, 1995; Sapkota et al., 2009).

Stand structure parametres allow predictions of forest biomass and can provide spatial information on potential determinants of plant species distributions (Couteron et al., 2005). It also relates to the basal area of trees, density of trees, and densities of herbs, shrubs, climbers, saplings and seedlings. The tree basal area of 71.05m^2/ha is high and comparable to the reported range of various Indian tropical forests (Visalakshi, 1995; Sapkota et al., 2009) and slightly higher than the value reported from Monteverde of Costa Rica (62 m^2/ha, Nadkarni et al., 1995). High basal area is a characteristic feature of mature forest stand and serves as a reflection of high performance of the trees. It may also presuppose the development of an extensive root system for efficient nutrient absorption for tree growth implication for subordinate plants as the big trees suppress the growth of small plants by intercepting much of the solar radiation that might otherwise reach the forest floor. The Dipterocarpaceae had the highest basal area in the present study, followed by Combretaceae, Myrtaceae, Rubiaceae and Moraceae. These families contain important timber species such as *Shorea robusta, Terminalia alata, Anogeissus latifolia, Syzygium cumini, Syzygium cerasoides, Terminalia bellirica, Terminalia chebula*, etc. The Barringtoniaceae, Chochlospermaceae, Clusiaceae, Malvaceae, Melastomataceae, Myrsinaceae, Ochnaceae, Rosaceae, Salicaceae, Lauraceae, Rhamnaceae, Sterculiaceae, Symplocaceae, Verbenaceae, Flacourtiaceae, and Rutaceae did not contribute much to the total basal area. In the biosphere reserve, Chochlospermaceae, Sapotaceae, Salicaceae, etc. was represented by one individual while Barringtoniaceae was represented by two individuals and, Clusiaceae and Melastomataceae were represented by three individuals each. This implies that very low contributions of these families to the total basal area may be due to their low numbers. Thus, these families may not be very important in terms of dominance. Species wise density of individuals having ≥30cm girth of the reserve ranged from less than one plant per hectare to 284 plants/ha and the total density of the reserve was 794 plants/ha. Maximum density (per hectare of individuals of ≥30cm) was recorded for *Shorea robusta* (284) followed by *Terminalia alata* (50), *Anogeissus latifolia* (45), *Protium serratum* (32) and *Dillenia pentagyna* (29). Density was observed less than or equal to one for many species like *Antidesma acidum, Artocarpus lacuccha, Butea monosperma,, Casearia elliptica, Chionanthus intermedicus, Cochlospermum religiosum, Euonymus glaber*, etc. All other species showed intermediate range of density per hectare. The tree density of 794 individuals/ha recorded for the reserve is lower as compared to densities reported from Saddle Peak of North Andaman Islands and Great Andaman Groups (946-1137 trees/ha, Padalia et al.,2004). However, the tree density is comparable with other tropical forests, e.g., Kalkad Western Ghats (575-855 trees/ha, Parthasarathy, 1999), Brazil (420-777 trees/ha, Campbell et al., 1992), seasonally deciduous forest of Central Brazil (734 trees/ha, Felfili et al., 2007), Semideciduous forest of Piracicaba, Brazil (842 trees/ha, Viana and Tabanez, 1996) and Costa Rica (617 trees/ha, Heaney and Proctor, 1990). There appears to be little literature available to compare the herb, shrub, sapling and seedling densities with at the local level. The reported densities of these vegetation layers of the reserve is well comparable to Mishra et al. (2008), the study conducted in some sites of the same biosphere reserve. The fewer numbers of saplings recorded in relation to seedlings implies that most of the saplings are transiting into young trees. It could also mean that most of the seedlings probably die due to intense competition (Weidelt, 1988) for available resources before they reach the sapling stage. Nevertheless, the totality of saplings and seedlings is colossal and reflects high regeneration potential of the forest (Mohanty et al., 2005; Khumbongmayum et al., 2006).

Diversity Measures

Species diversity, concentration of dominance and some mathematical indices of different vegetational layers of the reserve are given in Table -4.9. Measurement of biodiversity of specific area (local scale) on the basis of species richness does not provide a complete understanding about the individuals of the species in an ecosystem as it suffers from the lack of evenness or equitability. It was observed that the richness index ranged from 3.36 to 6.59 (tree layer), 0.55 to 2.22 (climber layer), 1.24 to 4.24 (herb layer), 1.66 to 2.92 (shrub layer), 2.98 to 6.15 (sapling layer) and 0.73 to 4.37 (seedling layer) and equitability showed little variation across the vegetational layers which ranged from 0.61 to 0.95 (tree layer), 0.8 to 0.93 (climber layer), 0.63 to 0.91 (herb layer), 0.76 to 0.96 (shrub layer), 0.82 to 0.95 (sapling layer) and 0.61 to 0.88 (seedling layer). Shannon Wiener's index of diversity is one of the popular measures of species diversity. It ranged from 1.80 to 3.11, 0.63 to 1.86, 1.76 to 2.37, 1.57 to 2.99, 1.01 to 2.62 and 2.1 to 3.03 for tree, climber, shrub, herb, seedling and sapling layers, respectively, across all sites. Maximum range of species diversity of 1.8 to 3.11 was experienced by tree layer and the minimum range of 0.63 to 1.86 by climber layer indicating that tree layer of SBR was highly diverse and climber layer was the least (Table-4.9).

Table 4.9: Species diversity (SD), Concentration of dominance (CD) species richness (SR) and species evenness (SE) of different forest strata of Similipal biosphere reserve.

Plant group	Range of diversity indices			
	SD	CD	SR	SE
Tree	1.8 -3.11	0.07 - 0.316	3.36 - 6.59	0.611 - 0.951
Climber	0.63 - 1.86	0.155 - 0.36	0.55 - 2.22	0.8 - 0.931
Shrub	1.76 - 2.37	0.102 - 0.216	1.66 - 2.92	0.76 - 0.96
Herb	1.57 - 2.99	0.053 - 0.323	1.24 - 4.24	0.63 - 0.91
Sapling	2.1 - 3.03	0.061 - 0.194	2.98 - 6.15	0.816 - 0.949
Seedling	1.01 - 2.62	0.129 - 0.397	0.73 - 4.37	0.61 - 0.88

The range of evenness value and Simpson's diversity index of 0.61-0.96 and 0.053-0.397, respectively in vegetation layers of Similipal imply that most of the species were equitably distributed among the species, while very few species showed the degree of dominance (Pascal and Pellissier, 1996). Shannon Wienner species diversity value among vegetational layers of the reserve ranged from 0.63-3.11 indicating that SBR was highly diverse. The species diversity is generally higher for tropical forests, which is reported as 5.06 and 5.40 for young and old stand, respectively (Knight, 1975). For Indian forests the diversity index ranges between 0.83- 4.1 (Visalakshi, 1995). The diversity index of different vegetational layers of SBR is well within the reported range for the forests of Indian sub-continent (Table-4.9). Higher species diversity index in tropical forests as reported by Knight (1975) in comparison to the present investigation may be due to differences in the area sampled and lack of uniform plot dimensions.

On the other hand the value obtained for the concentration of dominance for vegetation layers of SBR (0.053-0.397) was greater than those recorded in Nelliampathy (0.085; Chandrashekara and Ramakrishnan, 1994) and tropical dry deciduous forests of Western India (0.08- 0.16; Nirmal Kumar et al., 2010). The high dominance value in the present study indicates single species dominance by *Shorea robusta* in tree, sapling and seedling layers of the reserve (Table- 4.1, 4.5 and 4.6).

Comparative Analysis of Tree Species Diversity Under Various Tropical Forests

The tree diversity observed under various tropical forests has compared with the Similipal biosphere reserve and vice-versa (Table-4.10). Species richness and density of tree species of the reserve i.e. 117 species and 794 plants per hectare is well within the reported range of tropical forests in India and outside India. However, the basal area estimated for tree species in the present investigation is well within the reported range of Indian tropical forests but higher than that of tropical forests found outside India (Table-4.10). The high basal area of 71 m^2/ha obtained in the present investigation was largely due to the contribution of the dominant tree species of the biosphere reserve i.e. *Shorea robusta* which alone scored 39% (27.73 m^2/ha) of basal area.

Table 4.10: Number of species, genera, families, density (Plants/ha) and basal area (m^2/ha) of tree species at various tropical forests

Forest and location	No. of species	No. of genera	No. of families	Density (Plants/ha)	Basal area (m^2/ha)	Source
Indian tropical forest						
Moist deciduous forest, Similipal	117	87	42	793.67	71.04	Mishra et al., 2012
Moist deciduous forest, Andaman	235	153	73	946	28.60	Padalia et al., 2004
Semi evergreen forest, Andaman	231	153	71	1027	33.76	Padalia et al., 2004
Evergreen forest, Andaman	264	176	81	1137	44.28	Padalia et al., 2004
Wet evergreen forest, South western ghat	122	89	41	575-855	61.7-94.64	Parthasarathy, 1999
Tropical forests outside India						
Neotropical cloud forest, Monteverde, Costarica	114	83	47	555	62.0	Nadkarni et al., 1995
Seasonally deciduous forest, Iaciara, Brazil	39	-	-	734	16.73	Felfili et al, 2007
Seasonally deciduous forest, Monte Alegre, Brazil	56	-	-	633	19.36	Nascimento et al., 2004
Semideciduous, Piracicaba, SP, Brazil	101	-	-	842	12.53	Viana and Tabanez, 1996
Evergreen rain forest, Ngovayang (Cameroon)	99-121	-	-	451-634	28.8-42.1	Christelle et al., 2011
Subtropical forest, Bagh district, Kashmir, Pakistan	72	-	31	344	69.31	Saheen et al., 2011
African wet tropical forest	344-494	-	-	371-486	27.8-35.8	Chuyong et al., 2011

The overall analysis on phytosociological attributes of various vegetational layers indicates that species rich communities of the moist deciduous tropical forests are not only being reduced in area but they are also becoming species poor and less diverse due to rapid deforestation and forest fragmentation. The community organization is also changing in response to increased anthropogenic disturbance. The study has shown that SBR is highly rich in plant diversity and is one of the treasure houses of good ecological wealth of Eastern Ghat, India. The long history of timber exploitations prior to its conversion into a biosphere reserve has resulted in the alteration of structure of the forest whereby most of the tree species were represented by very few individuals. The ecological importance of most of the tree species was also low, which reflected rarity of most of the species. However, the abundance of small trees coupled with the colossal sum of saplings and seedlings reflects a high regeneration potential of the forest.

The forest management issues associated of the Similipal Biosphere Reserve (SBR) could be addressed by collection and analysis of long-term ecological data which requires scientific baseline studies. We have covered extensively structural parameter analysis which is helpful to know the present state of ecological health of the ecosystem. But due to the various forms of anthropogenic pressures to destroy habitat for logging, illegal hunting, and other challenges (mining in periphery, etc.), the conservation efforts may not yield desired result. With continued biotic pressure and consequent change in structure and function of ecosystem, our management methodology also needs to be modified which can be achieved only by developing Long Term Research Network (LTRN). Similipal is globally recognized ecosystem covered under UNESCO's Biosphere Reserve housing wide range of flora and fauna. We need to carry out research and education activities to create an institutional platform to academicians, researchers and scientists. This ecosystem is under pressure. Continued destruction of old-growth and pristine forests of Similipal with high biodiversity will have a regional impact on social and ecological sustainability.

The overexploitation of natural resources in tropical world for meeting the basic needs of food, fodder and shelter of local population has disturbed the landscapes causing rapid depletion of biodiversity. Research results of workers on various ecological aspects of SBR may be of some help to develop management schemes for conservation of biodiversity. Lack of data base and structural and functional characters of the ecosystem at regular intervals will not help to develop a long-term strategy for sustainable development. Thus continuous collection of data as per long-term action plan on successional status of species level, external and local pressures on the ecosystem, soil fertility management and linkage between social and ecological processes is needed. The community participation and use of traditional ecological knowledge as tools for natural resource management should be integrated to achieve sustainable resource management and ecological rehabilitation.

Chapter-5

LITTER DECOMPOSITION AS AN ECOSYSTEM SERVICE

In deciduous forests, plants have the characteristics of shedding their leaves, twigs, flowers and fruits annually which is termed as the litter fall. It plays an important role in forest ecosystems in the way of the return of organic matter and nutrients from aerial parts of the plant community to the soil surface and enhance the soil fertility through decomposition of the forest floor litter (Knoepp *et al.*, 2005). The cycling of matter is inherent in ecosystem services (Barbour et al., 1980).

Not only it is an essential part of nutrient cycling, but also acts as a protective layer by buffering changes in soil water content, temperature and hindering erosion, leaching and soil compaction (Sayer, 2006). Such activities performed by the forest floor are responsible for maintaining the integrity of an ecosystem through transfer of matter and nutrients (Rajendraprasad *et al.*, 2000). In tropical ecosystems, maintenance of soil organic pool is achieved by high and rapid circulation of nutrients through the fall and decomposition of litter. Standing crop of litter (total forest floor material) acts as an input-output system of nutrients and the rates at which forest litter falls, and subsequently decays, regulate energy flow, primary productivity and nutrient cycling in forest ecosystems (Sunderpandian and Swamy, 1999).

Among different ecosystems services litter decomposition is a critical service that removes wastes, recycles nutrients, and renews soil fertility and carbon sequestration (Wall and Virginia, 2000). During the process of decomposition, dead organic matters convert into smaller and simpler compounds. The products of complete decomposition are carbon dioxide, water, and inorganic ions (like ammonium, nitrate, phosphate, and sulphate). In fact, the importance of litter decomposition as ecological services that supports other ecosystem services. For instance, biomass production, nutrient cycling and biodiversity conservation are a few among other ecological services for which the litter decomposition plays a central role. It may also be pointed out here that the forest ecologists have paid considerable attention to litter decomposition in relation to nutrient cycling and soil productivity. The obvious reason is that litter decay has a pronounced effect on the availability of nutrients, and nutrient availability is the basic determinant of biodiversity, plant growth and productivity.

Litter decomposition is mainly a biological process carried out by insects, worms, bacteria, and fungi, both on soil surface and in the soil. However, apart from soil micro and macro faunal activity there are four more factors affecting litter decomposition. They include (a) climatic factors, (b) substrate and its quality, (c) type of vegetation, and (d) vegetation cover. In this chapter discussion was made on methods to study litter decomposition, litter depth influencing plant diversity, the factors influencing litter decomposition, rate of litter decomposition and the pattern of nutrient nutrient release to the forest floor of Similipal.

Methods to Study Litter Decomposition

Litter decomposition is most commonly measured using the litter bag technique. A known quantity of litter is placed into a mesh bag which is then inserted into the litter layer of a forest floor. Bags are harvested at periodic intervals, dried and reweighed to determine the amount of mass lost. By incubating the litter in situ, they are exposed to the normal fluctuation in temperature and moisture. The litter bag technique has both advantages and disadvantages (Woods and Raison, 1982). The advantage of this technique is it allows registering the litter weight loss in field and the subsequent chemical and biological examination of the material involved (Weber, 1987). Typically, 1-2 mm mesh size and 10×10cm to 30×30cm nylon bags or fibre glass screen or polyvinyl bags with 5-20 g (dry weight) samples are used (Anderson and Ingram, 1993). However, the mesh-size may hinder soil faunal activities. In this context, improved methods for litter decomposition studies need to be developed. The decomposition rate constant, k, is calculated from the decay curve using the following equation

$$\ln (M_0/M_t) = k \times t$$

Where M_0 = mass of litter at time 0, M_t =mass of litter at time t, t= time of incubation, and k = decomposition rate constant.

Half lives ($t_{0.5}$) of decomposing litter samples are estimated from the values as:

$t_{0.5} = 0.693/k$

Similarly, time taken for 95% decay estimated as follows

$t_{0.95} = 2.9957/k$

According to Mc Claugherty and Berg (1987) the single exponential model described above may be suitable for homogeneous substrate, and the materials with high nutritional status especially nitrogen and complex organic constituents such as lignin and tannins. However, double exponential model (Bunnell and Tait, 1974) may be suitable for analysing the decomposition data obtained for heterogeneous substrate.

Rate of decomposition can also be measured by calculating the ratio of annual litter fall to the equilibrium litter on the forest floor. Generally, the equilibrium (or the steady state) litter on the plots are measured biannually in two or more successive years. The average annual litter fall measured by adopting litter bag technique. According to Anderson and Swift (1983) the litter turnover coefficient calculated by this method is suitable for comparison of humid tropical forests. The other methods commonly used for litter decomposition studies include, (a) measurement of carbon dioxide evolution or oxygen uptake and (b) determination of the microbial biomass in the litter layer. The approach to determine the rate of decomposition should involve usage of both single exponential model and double exponential model.

Plant Diversity and Litter Depth

Plant diversity is influenced by litter depth (Grime, 1979) and the heterogeneous nature of litter cover on the forest floor promotes species coexistence by facilitating seedling emergence and influencing seedling survival through the forest (Facelli and Pickett, 1991 a). The quantity of litter fall varies greatly over a range of spatial and temporal scales and is determined mainly by climate, seasonality, topography, and soil parent materials and species distribution. Climate change may affect forest floor litter biomass as change in rainfall patterns and mean annual temperatures can also affect tree phenology and tree species distribution (Condit *et al.*, 1996; Mishra *et al.*, 2006) and increase in productivity (Zak *et al.*, 2003). The forest floor litter layer acts as an interface between the soil surface and the atmosphere, and provides a degree of protection to the soil surface by intercepting rain (Bankobi *et al.*, 1993) and solar radiation (Ogee and Brunet, 2002), and buffering the soil surface against fluctuations in temperature (Ponge *et al.*, 1993) and water content (Ginter *et al.*, 1979). Litter is also the major source of soil organic matter which strongly influences the structure of the soil and increases its stability (Marshall *et al.*, 1996). Soil organic matter often plays a role in determining the pH of the soil surface horizons, as p^{H} is regulated by humic acids to a large extent in some soils (Wilke *et al.*, 1993). Litter and soil organic matter affect soil porosity and aeration indirectly, as constitute the major food supply to the soil biota (Gonzalez and Zou, 1999).

Forest floor litter biomass directly or indirectly influence plant growth and survival. Changes to the soil water balance and the nutrient cycle constitute direct influences on plant growth as they modify the availability of resources. Soil compaction and other changes to the soil structure indirectly affect tree growth by diminishing the available pore space, thus inhibiting root expansion (Sayer, 2006). Increase or decrease in organic matter content can cause changes in p^H, base saturation and the water retention capacity of the soil (Marshall *et al.,* 1996), all of which may influence nutrient uptake, and therefore potentially growth and survival of the vegetation. Furthermore, forest floor litter can influence seed germination and seedling establishment and survival both positively and negatively, changing competitive outcomes between tree seedlings and forbs (Barritt and Facelli, 2001) and thus contributing to the species composition and structure of the forest (Facelli and Pickett, 1991b). Naturally occurring patchiness of litter and differing litter depths and structure facilitate the establishment of seedlings of some species while inhibiting others and thus contribute to small scale heterogeneity of the vegetation and coexistence of species (Dzwonko and Gawronski, 2002 a, b). Canopy cover of tree species composition accounts forest floor litter biomass and show considerable variation. This spatial heterogeneity allows coexistence of species and contributes to niche differentiation for seedlings and forest floor microorganisms (Sayer, 2006). To further our knowledge on the ecosystem functioning of the forest ecosystems of Similipal Biosphere Reserve (SBR) categorisation of forest floor litter was made quantitatively along with the turn over rate and turn over time of the forest floor litter in relation to climate, and the change in vegetation dynamics and soil physicochemical characteristics in relation to forest floor litter biomass.

Forest Floor Biomass

Monthly litter fall values at different sites of SBR are given in Fig. 5.1. At all the sites, litter biomass started increasing from the month of October and attained maximum value in March. However at some of the sites (Lulung, Handipuhan and Podadhia) the maximum litter biomass was marked in the month of December. Low rates of litter fall are consistently recorded in August-September (Fig. 5.1). This trend indicates that litter fall rate may vary according to age, structure and type of vegetation, especially of the upper canopy even though generally high amounts of litter input occur during the dry season and low amounts in the wet season. The maximum litter fall biomass is accumulated at Nigirdha (583.92 g/m^2/yr) followed by Joranda, Jenabil, Bhanjabasa, Chahala, Gurguria, Kalikaprasad, Lulung, Padadiha and Handipuhan (Fig. 5.2). The mean annual forest floor litter biomass obtained from different study sites of SBR varied from 304 to 584 g/m^2/yr, lie within the range reported from dry deciduous tropical forests of India (Rai and Srivastava, 1982; Rajendra Prasad *et al.,* 2000). Seasonal accumulation rate of total litter biomass across all study sites was in the order of winter > summer > rainy.

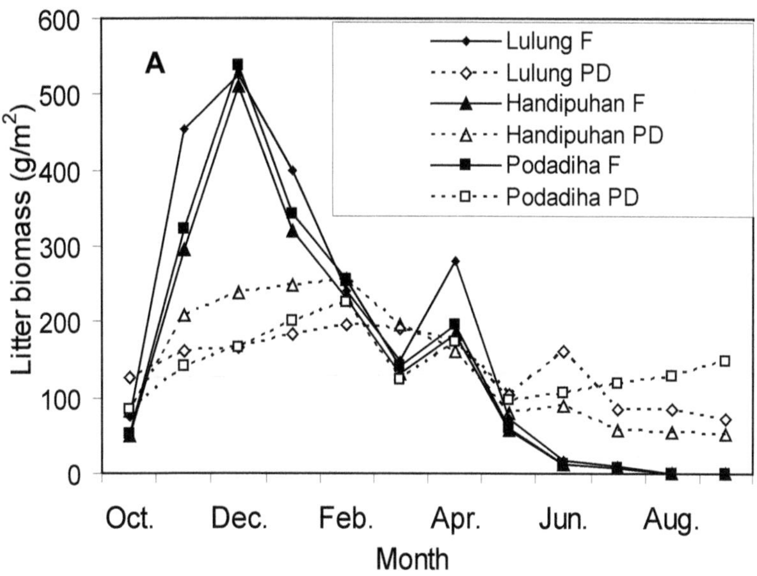
A
Lulung F
Lulung PD
Handipuhan F
Handipuhan PD
Podadiha F
Podadiha PD
Litter biomass (g/m2)
0
100
200
300
400
500
600
Oct.
Dec.
Feb.
Apr.
Jun.
Aug.
Month

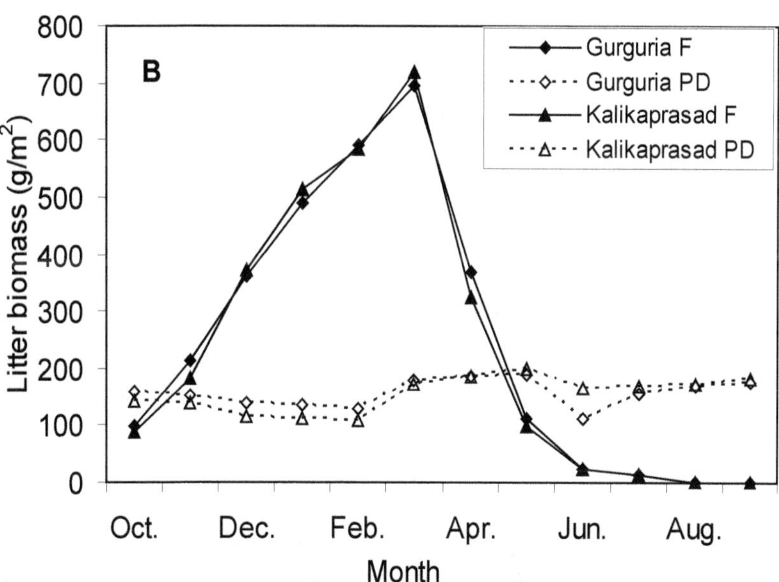
B
Gurguria F
Gurguria PD
Kalikaprasad F
Kalikaprasad PD
Litter biomass (g/m2)
0
100
200
300
400
500
600
700
800
Oct.
Dec.
Feb.
Apr.
Jun.
Aug.
Month

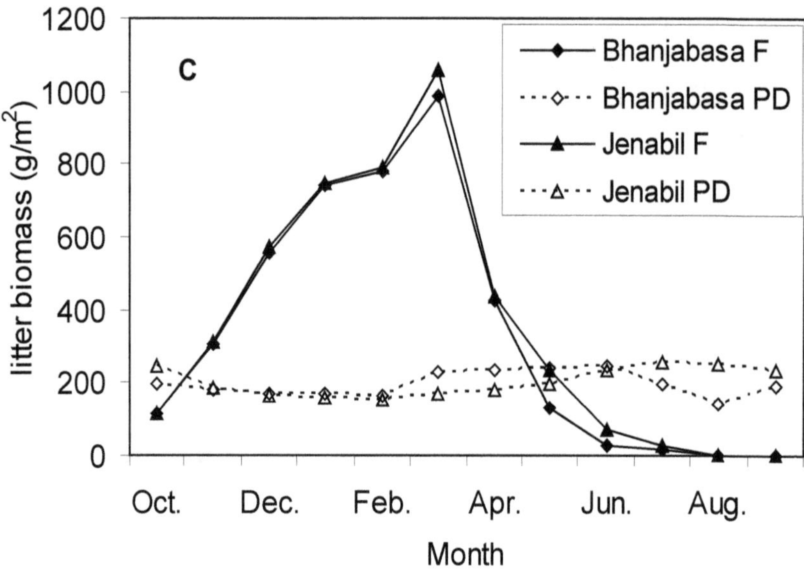

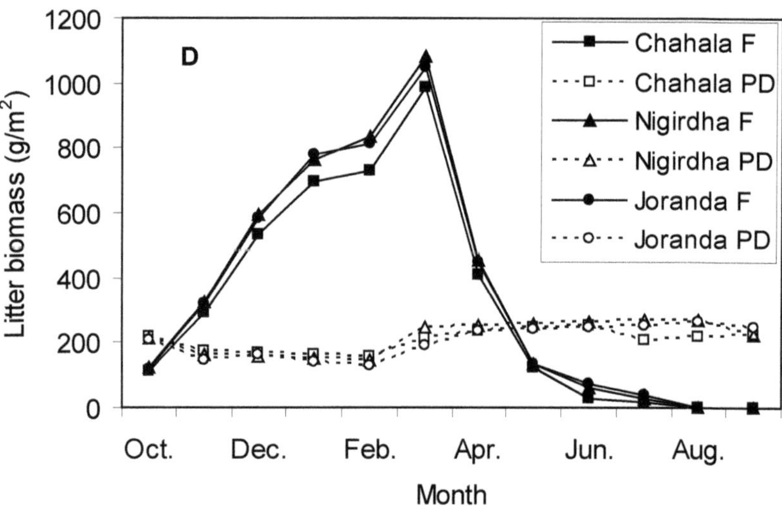

Figure 5.1: Variation in monthly forest floor litter biomass (F = fresh and PD = partly decomposed) at transition (A), buffer (B), south core (C) and north core (D) sites of Similipal biosphere reserve.

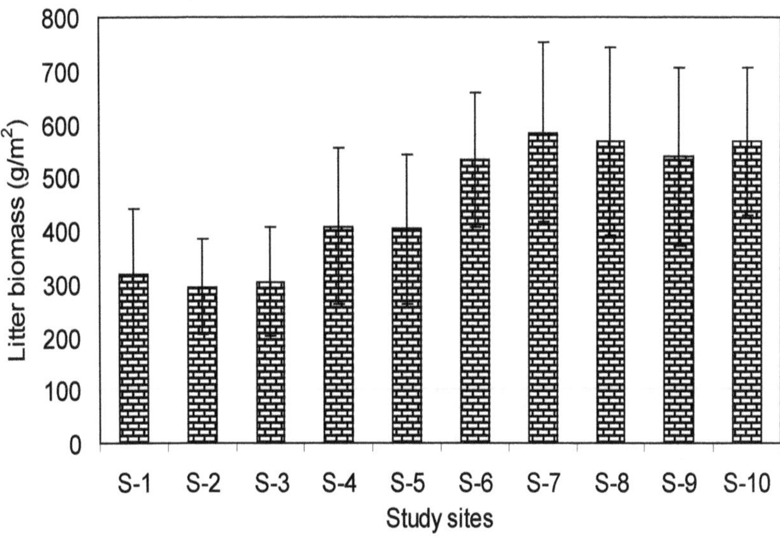

Figure 5.2: Average yearly forest floor litter biomass at different study sites of Similipal biosphere reserve. (S-1: Lulung; S-2: Handipuhan; S-3: Podadiha; S-4:Gurguria; S-5: Kalikaprasad; S-6: Chahala; S-7: Nigirdha; S-8: Joranda; S-9: Bhanjabasa; S-10: Jenabil).

Forest floor litter was categorized into two components, i.e., fresh and partly decomposed litter. Seasonal accumulation trend of fresh litter biomass follows the same trend as that of total forest floor biomass. While in case of partly decomposed biomass seasonal accumulation trend was in three different orders as:

(i). Winter > summer > rainy at Lulung and Handipuhan and Podadiha;

(ii). Summer > rainy > winter at Kalikaprasad, Chahala, Bhanjabasa and Nigirdha; and

(iii).Rainy > summer > winter at Gurguria, Jenabil and Joranda (Fig. 5.1).

Most of the study sites recorded lesser quantities of partly decomposed litter biomass in rainy season compared to summer and winter seasons. The sites located in transition areas also received less litter fall than the core and buffer areas of the reserve. Two peaks of fresh litter input were marked in all the study sites of the reserve. In the transition area the peak was observed in December and the second one in April while in the core and buffer areas of the reserve first peak was observed in March and the second one was in May. However, maximum fresh litter input in transition areas was recorded in the month of December whereas maximum litter input occurred in March both in core and buffer areas of the reserve. Analysis of variance indicated significant difference in forest floor biomass between the sites, seasons and category. All these have also significant interactive effect on forest floor litter biomass (Table-5.1). At all the study sites the maximum litter fall biomass was recorded in drier part of the year and the lowest during the wetter part of the year (Fig. 5.1). This behaviour indicates that most of the upper canopy tree species are dry deciduous type and very few of them are ever green and semi-evergreen type (Rajendraprasad *et al.*, 2006).

Rate and Time of Forest Floor Biomass

Litter breakdown and mineralisation is also governed by soil and forest floor micro and macro-organisms and its rate is regulated by the interaction between these organisms and the litter resources (Anderson and Swift, 1983; Upadhyay and Singh, 1989; Upadhyay *et al.*, 1989; Rajendraprasad, 1995). In the present study low turn-over coefficient was observed during the monsoon period (July to October, the wet months in Odisha) followed by an increase in summer and winter seasons (Table-5.2). Site to site variation in the litter turnover was also marked in different seasons of a year (Table-5.3).

Table 5.1: Analysis of variance (ANOVA) of litter production (Litter biomass), turnover rate (k) and turnover time (t) of Similipal biosphere reserve (SBR).

Source		Degree of freedom	F- ratio	Significance Level
Total litter biomass	Treatment	30	7.44	P < 0.001
	Season	02	44.28	P < 0.001
	Site	09	3.72	P < 0.001
	Category	01	25.21	P < 0.001
	Season X Site X Category	18	7.11	P < 0.001
	Error	209	-	
Turnover rate (k)	Treatment	29	4.5	P < 0.001
	Season	2	5.75	P < 0.01
	Site	9	15.1	P < 0.001
	Season X site	18	7.0	P < 0.001
	Error	60	-	
Turnover time (t)	Treatment	29	7.449	P < 0.001
	Season	2	10.761	P < 0.001
	Site	9	7.393	P < 0.001
	Season X site	18	7.111	P < 0.001
	Error	60	-	

Source: Mishra et al. (2007).

Table 5.2: Seasonal variation in the turnover coefficient (k) at different study sites of Similipal biosphere reserve (SBR).

Site	Winter	Summer	Rainy
Lulung	0.83± 0.14	0.61 ± 0.13	0.41 ± 0.02
Handipuhan	0.78 ± 0.11	0.52 ± 0.09	0.44 ± 0.02
Podadiha	0.72 ± 0.12	0.51 ± 0.11	0.46 ± 0.025
Gurguria	0.89 ± 0.16	0.61 ± 0.14	0.44 ± 0.019
Kalikaprasad	0.86 ± 0.13	0.63 ± 0.15	0.42 ± 0.01
Chahala	0.96 ± 0.17	0.65 ± 0.12	0.55 ± 0.015
Nigirdha	0.95 ± 0.21	0.65 ± 0.18	0.58 ± 0.023
Joranda	0.95 ± 0.2	0.63 ± 0.17	0.53 ± 0.018
Bhanjabasa	0.98 ± 0.12	0.71 ± 0.11	0.56 ± 0.013
Jenabil	0.96 ± 0.12	0.68 ± 0.16	0.56 ± 0.02

Source: Mishra et al., (2007)

Table 5.3: Seasonal variation in turnover-time (t, years) at different study sites of Similipal biosphere reserve (SBR).

Site	Winter	Summer	Rainy
Lulung	1.21± 0.33	1.64 ± 0.28	2.44 ± 0.61
Handipuhan	1.28 ± 0.25	1.92 ± 0.51	2.27 ± 0.83
Podadiha	1.39 ± 0.28	1.96 ± 0.64	2.17 ± 1.14
Gurguria	1.12 ± 0.30	1.64 ± 0.58	2.27 ± 0.73
Kalikaprasad	1.16 ± 0.27	1.59 ± 0.47	2.38 ± 1.12
Chahala	1.04 ± 0.21	1.56 ± 0.31	1.81 ± 1.15
Nigirdha	1.05 ± 0.18	1.54 ± 0.47	1.72 ± 1.16
Joranda	1.05 ± 0.14	1.59 ± 0.63	1.89 ± 0.87
Bhanjabasa	1.02 ± 0.23	1.41 ± 0.24	1.78 ± 1.31
Jenabil	1.04 ± 0.27	1.49 ± 0.51	1.78 ± 1.54

Source: Mishra et al. (2007)

The annual value of the turnover coefficient was maximum at Bhanjabasa (0.92) followed by Chahala = Jenabil = Nigirdha, Joranda, Kalikaprasad, Gurguria, Lulung, Handipuhan and Podadiha (Table-5.4). The turnover time for litter ranged from 1.1 to 1.45 years for complete decomposition of litter biomass of the forest floor of SBR (Table-5.4). Among the sites maximum turnover time was required at Podadiha and minimum at Bhanjabasa (Table-5.4).

Table 5.4: Variation in the litter turnover coefficient (k) and litter turnover time (t) of the study sites of Similipal biosphere reserve (SBR).

Site	Turnover coefficient	Turn over time (t)
Lulung	0.82	1.22
Handipuhan	0.71	1.41
Podadiha	0.69	1.45
Gurguria	0.83	1.21
Kalikaprasad	0.85	1.18
Chahala	0.89	1.12
Nigirdha	0.89	1.12
Joranda	0.86	1.16
Bhanjabasa	0.92	1.1
Jenabil	0.89	1.12

Source: Mishra et al. (2007)

The variation observed in case of turnover rate and turnover time was significant between the sites and seasons. The site and season both have significant interactive effect of turnover rate and turnover time of the forest floor litter biomass of Similipal. In the tropical climate of Similipal temperature and humidity are moderate throughout the year. Rainfall is abundant with rains for nearly 5 to 6 months in a year. But the major precipitation occurs between July and October and other rainy days are distributed over the rest of the months (Mishra *et al.*, 2006). According to Xiong and Nilsson (1997) such conditions are favourable for microbial activities except during certain months. The investigation also emphasis during the heavy rain period the decomposition rate was low and this could be explained in as due to the continuous heavy rain, the soil becomes supersaturated with water accompanied by low oxygen content and low light intensity, resulting in retarded activity of the micro flora and microbes. Altering wet and dry periods could also be important, since Polunin (1984) suggested that the higher the frequency of change between wetting and drying, the greater the potential release of nutrients from the litter. Rajendraprasad *et al.* (2000) suggested that the aerobic digestion of cellulose is inefficient during active rainy season but becomes active during dry season due to the enhanced activity of microorganisms. These reasons may, therefore, account for the low turnover rate of litter during the rainy season and the high turnover rate of litter during the dry season and at the onset of periodical and interrupted summer rains during the months of April-May in the SBR.

Factors Affecting Litter Decomposition

The rate of patterns of litter decomposition are dependent on the interaction of climate, soil biota and quality and quantity of organic matter (Swift et al., 1979). One can predict gross estimates of decomposition based on the climate and the C:N:lignin rations of organic matter (litter). The primary factors which affect litter decomposition are discussed under following heads: (a) climate, (b) vegetation, (c) substrate and its quality and (d) soil biota.

Climatic Factors

Climate markedly modifies the nature and rapidity of litter decomposition. Moisture and temperature are among the most crucial variables (Brinson, 1977) because they affect both the development of plant cover and the activities of microorganisms which are highly critical factors in soil formation. Effects of soil moisture on litter decomposition are little complicated. Decomposition is inhibited in very dry condition. Decomposition is also slow in very wet soils because anaerobic condition develops in saturated soils. Decomposition proceeds at faster rate at intermediate water contents. Kononova (1975), citing several other publications, concluded that the highest intensity of organic matter decomposition was observed when the soil moisture content of about 60-80 % of its maximum water-holding capacity. According to Van der Drift (1963) moisture passing through the detritus may be important in speeding decomption. Therefore, studies in litter decay should not be compared unless moisture regimes are the same.Temperature is often the primary factor determining rates of litter decomposition (meentemeyer, 1978, Anderson, 1991, Hobbie, 1996) and decomposition rate are generally more sensitive to temperature than are rates of net primary production (Lloid and Taylor, 1994). Thus, the balance between ecosystem C fixation and decomposition may be altered under a warmer climate, potentially causing a dramatic increase in the flux of CO_2 from soils to the atmosphere (Cox et al., 2000). For each 10°C increase in temperature between 20 and 40°C the rate of CO_2 production doubled (Wiant, 1967). No CO_2 production at all was detected at 10°C and 50°C or above it declined markedly. According to Kononova (1975) the highest intensity of organic matter decomposition was observed under conditions of moderate temperature (about 30° C). Increase or decrease of temperature beyond the optimal levels brought about a decline in the rate of organic matter decomposition. Therefore, studies on litter decay should not be compared unless temperature regimes are the same. Differences in litter decomposition rate at various altitudes, due to variation in temperature were reported by William and Gray (1974). Shanks and Olson (1961) compared litter decay beneath natural stands at various elevations and concluded that there was average decrease in breakdown of nearly 2% for each 1°C drop in mean temperature. The influence of temperature on the decomposition of lignin is especially marked. At 37°C, lignin decomposes rapidly, with 50 – 60% of it disappearing within 9 months (Waksman and Gerretsen, 1931). Meentemeyer (1978) used annual actual evapotranspiration as the index of predictor variable of decomposition rate.

The decomposition rates in different types of forests in the SBR appear to be correlated with rainfall. According to Swamy (1989) the rate of decomposition in rainfall rich forests is faster than in those where the rainfall is comparatively less. The percolating water from rainfall may leach the excrements and remains of organisms down to the lower horizons, where other specialized microbes will attack the remaining organic matter (Van der Drift, 1963).

The litter breakdown rate varies with season. Gholz et al. (2000) and Loomis (1975) found that decomposition was rapid in summer, whereas Lang (1974) estimates the leaf litter decay to 3.75 $g/m^2/day$ during the autumn months and

0.80 g/m^2/day during the remainder of the year. Boonyawat and Ngamponsai (1974) supported Lang's result when they found that the highest decomposition of evergreen forest litter occurred in the late rainy season and early winter (0.36 t / ha/month Table-5.5) and the lowest rate in summer (0.14 t/ha/month).

Effects of Vegetation on Forest Floor Biomass

Vegetation composition, profile and canopy cover of forest ecosystems influencing the quantification of forest floor litter biomass. Not only this but also seasonal variation in litter production is influenced by plant physiognomic groups. From the investigation on SBR it was noticed that maximum forest floor litter biomass was collected during the dry seasons of the year in comparison to the wet season (Table-5.6). This is only due to the seasonality of litter fall. Inside the reserve most of the upper canopy trees are deciduous in nature and they shed their leaves towards the end of the winter and beginning of the summer (Mishra *et al.*, 2006). However, difference in forest floor litter biomass among the study sites distributed in three different zones, i.e., core, buffer and transition of the reserve may be due to the difference in above ground phytomass, species diversity, tree density and structure of vegetation (Table-5.7). Rajendraprasad *et al.* (2000) and, Pragasan also reported the same type of phenomenon in sacred groves of Kerala and Coromandal coast of south India, respectively. Linear regression analysis taking litter fall as a function of tree density, basal area, bio-volume, canopy and species diversity showed significant correlation while significant negative correlation showed with the richness of herbs, shrubs and climbers (Table-5.8). Table-5.8 shows coefficient of correlation (r) and regression parametres of litter biomass with biotic (density, basal area, bio-volume) and abiotic components of SBR. A similar relationship between tree density, basal area and litter production has been reported for tropical forests (Martinez and Sarukhan, 1990; Rajendraprasad *et al.*, 2000).

Table 5.5: Analysis of variance (ANOVA) table for testing the significance of multiple correlation coefficient (r) of litter biomass with rainfall, temperature and humidity of Similipal biosphere reserve (SBR).

Source	Df.	SS	MSS	F-ratio (df.)
Regression SS	3	719106.22	239702.1	16.312 (3,8)
Error	8	117559.31	14694.9	
D.f. (Difference) at P < 0.005 = 9.60 (Tabulated "F") Calculated "F" = 16.312 So, it is significant at P < 0.05				

Source: Mishra et al. (2007)

Table 5.6: Seasonal variation in total litter biomass (g/m^2) at different study sites of Simlipal biosphere reserve.

Site	Summer	Winter	Rainy
Lulung	291±66	556±107	110±39
Handipuhan	229±47	579±120	74±37
Podadiha	229±42	549±110	134±17
Gurguria	472±122	576±155	182±43
Kalikaprasad	473±111	547±165	186±26
Chahala	626±111	729±152	243±52
Nigirdha	692±158	782±175	278±46
Joranda	654±123	769±221	281±36
Bhanjabasa	636±127	779±172	201±56
Jenabil	650±111	773±132	278±56

Source: Mishra et al. (2007)

Forest Floor Biomass and Soil Physical Characters

Soil surface covered with litter and living vegetation protects the soil from the direct impact of rain drops and through-fall (Sayer, 2006). Litter cover also decreases or prevents the destruction aggregates and separation of fine particles by rain drop impact, prevents soil surface compaction and sealing and consequently helps to prevent run-off and subsequent erosion during rainfall events (Marshall *et al.*, 1996). Besides such activities the forest floor litter also responsible for changing the soil physical properties such as bulk density, pore space, etc. From the investigations it was found that bulk density of soil in three different zones of the SBR is in the order of transition > buffer > core and also reflecting higher bulk density at low forest floor litter biomass sites in comparison to the higher forest floor litter biomass sites. However, the reverse trend was noticed for soil pore space and moisture content (Table-5.9). Linear regression analysis between forest floor biomass with physical properties of soil showed significant positive correlation between forest floor biomass with soil moisture content while negative between forest floor biomass with bulk density (Table-5.10). Similar type of observation is marked by Sayer (2006) in different forest ecosystems of central Europe.

Table- 5.7: Floristic wealth and diversity of different study sites of Simlipal biosphere reserve.

Name of the study site	No. of tree Species	No. of shrub species	No. of liana species	No of herb species	Density of tree sp. (Plants/ha)	Basal area of tree sp. (m^2/ha)	Bio-Volume (m^3/ha)	Canopy area (m^2/ha)	Sp. Div. (H^1)	Conc. of dominate (Cd)
Lulung	29	31	6	40	725	60	223	902	2.153	0.206
Handipuhan	20	33	4	48	715	49	196	789	2.426	0.131
Podadiha	29	29	4	45	680	52	218	864	2.139	0.103
Gurguria	22	26	5	34	740	63	338	1026	2.133	0.254
Kalikaprasad	23	23	3	36	750	79	333	1060	1.994	0.313
Chahala	36	18	2	28	875	89	447	1380	3.083	0.07
Nigirdha	27	16	3	29	985	85	449	1637	2.935	0.188
Joranda	31	20	2	26	810	81	397	1346	2.896	0.193
Bhanjabasa	22	19	3	30	950	86	439	1844	2.209	0.228
Jenabil	31	156	4	29	895	95	538	1858	3.107	0.069

Source: Mishra et al. (2008)

Table- 5.8: Relationship between forest floor biomass (FFB) and phytosociological characters.

Parametres	Correlation co-efficient 'r'	Intercept	Slope	Probability
Tree dens.Vs FFB	0.8	459.45	0.729	P<0.01
Basal area Vs FFB	0.91	12.5	0.13	P<0.001
Bio-volume Vs FFB	0.92	– 77.92	0.93	P<0.001
Tree species richness Vs FFB	0.4	19.23	0.002	NS
Shrub species rich ness Vs FFB	– 0.96	46.47	– 0.004	P<0.001
Climber species richness Vs FFB	– 0.65	6.71	– 0.0006	P<0.05
Herb species richness Vs FFB	– 0.95	61.95	– 0.005	P<0.001
Species div. Vs FFB	0.73	1.271	0.0002	P<0.01
Canopy cover Vs FFB	0.91	–125.55	0.257	P<0.001

Dens. : Density, FFB: Forest Floor Biomass, Bulk dens.: Bulk density, NS: Not significant.

Source: Mishra et al. (2007)

Table 5.9: Physical properties of soil across the study sites in Similipal biosphere reserve (Mean ±SEM)

Site	Bulk density (g/m3)	% of pore space	% of moisture content
Lulung	1.279 ± 0.092	54.31 ± 1.177	9.872 ± 0.449
Handipuhan	1.103 ± 0.139	58.82 ± 2.217	5.978 ± 0.093
Podadiha	1.452 ± 0.064	58.36 ± 1.274	5.856 ± 0.695
Gurguria	1.589 ± 0.038	47.727 ± 0.984	6.362 ± 0.469
Kalikaprasad	0.949 ± 0.037	70.297 ± 2.594	7.257 ± 0.854
Chahala	0.799 ± 0.023	59.413 ± 1.265	27.822 ± 0.947
Nigirdha	0.802 ± 0.007	72.325 ± 1.865	26.327 ± 1.084
Joranda	0.823 ± 0.031	56.337 ± 2.298	30.117 ± 0.824
Bhanjabasa	0.898 ± 0.008	73.22 ± 2.563	34.917 ± 1.329
Jenabil	0.912 ± 0.067	75.887 ± 2.751	33.112 ± 2097

Source: Mishra et al. (2007)

Table-5.10: Coefficient of correlation (r) and regression parameters of litter biomass with soil physical parameters of Similipal biosphere reserve.

Parameter	Correlation co-efficient 'r'	Intercept	Slope	Probability
Bulk dens. Vs FFB	- 0.789	9433.59	3965.95	P<0.05
Moisture Vs FFB	0.873	3162.34	104.69	P<0.001
Pore space Vs FFB	0.451	944.64	67.52	NS

FFB: Forest Floor Biomass, Bulk dens.: Bulk density, NS: Not signi cant

Table-5.11: Chemical properties of soil across the study sites in Similipal biosphere reserve (Mean +SEM)

Site	Organic Carbon (%)	Nitrate Nitrogen (Kg/ha)	Available phosphorus (Kg/ ha)
Lulung	0.482 ± 0.011	8.072± 0.471	4.528± 0.229
Handipuhan	0.453± 0.010	5.956± 0.327	19.710± 0.549
Podadiha	0.524± 0.012	5.839±0.826	3.414± 0.199
Gurguria	0.618± 0.010	7.823± 0.737	8.782± 0.349
Kalikaprasad	0.579± 0.013	7.538± 0.519	11.38±0.354
Chahala	0.650± 0.026	28.978± 2.497	17.678± 0.616
Nigirdha	1.10± 0.014	17.558± 1.842	23.334± 0.625
Joranda	0.612± 0.016	16.640± 2.435	28.118±0.719
Bhanjabasa	0.759± 0.011	27.575± 2.102	21.922±0.639
Jenabil	0.835± 0.011	28.595± 2.207	8.922±0.473

Table-5.12: Coefficient of correlation (r) and regression parameters of litter biomass with chemical properties of soil in Similipal biosphere reserve.

Parameter	Correlation co-efficient 'r'	Intercept	Slope	Probability
FFB with soil organic carbon	0.792	0.079	0.0001	P<0.01
FFB with nitrate nitrogen	0.83	-15.49	0.006	P<0.01
FFB with available phosphorus	0.61	-4.99	0.004	P<0.05

FFB: Forest Floor Biomass.

Forest Floor Biomass and Soil Chemical Characters

Forest floor litter biomass an inherent part of the soil nutrient status, which directly or indirectly regulate the growth and development of plants (Killian, 1998). In tropical forests litter removal is more frequent in comparison to any other forests. Thus in tropical forests deciduous trees are usually badly adapted to low nutrient conditions (Givinish, 2000). Dzwronski (2002a) stated that forest floor litter biomass and their decay affects the chemical properties of soil and their removal from the forest floor reduces the soil nutrient status and also regulating the leaching of nutrients into different depths of soil. From the result of the findings of forest floor biomass with respect to soil chemical characters it was observed that low forest floor biomass in transition and buffer areas of the reserve in comparison to the core areas of the reserve leads to maximum value of soil organic carbon, nitrate nitrogen and available phosphorus in the core in comparison to buffer and transition areas of the reserve. Significant positive correlation of forest floor biomass with soil organic carbon, nitrate nitrogen and available phosphorus indicates that sites having higher forest floor biomass had highest value of soil nutrient. Similar type of relationship between forest floor biomass and soil chemical nutrients is also marked by Mo *et al.* (2003) in tropical forests of China and Sayer (2006) in the forests of central Europe.

Soil Biota

Soil biota, primarily at a functional group level, is known to regulate ecosystem processes such as decomposition, carbon sequestration and nutrient cycling (Paustion et al., 2000). Soil biota, playing a meditative role, in decomposition may affect the type and availability of nutrients and thus community interactions. Evidence from the 1980's on deserts (Whitford et al., 1982) and from sub-alpine and wet and dry tropical ecosystems (Gonzalez and Seastedt, 2001) again indicate that soil fauna are key to litter decomposition. It is also reported that the seasonal variation in rate of decomposition could be variation in abundance of soil fauna. For instance, Madge (1965) concluded that since more animals are there on the litter layer during the wet season than in dry season, rate of decomposition was faster in the former season. When animals were completely excluded for nine months, no visible breakdown of oak and beech leaf litter occurred (Edwards and Heath, 1963). The same investigators reported that in earthworms removed litter the decomposition rate is three times faster than the litter from where smaller invertebrates such as springtails, enchytraeids and dipterous larvae were removed. A common carbomate insecticide (carbofuran), when applied at recommended dosaga, reduced the decomposition rate of red maple to between 0.99 and 1.26 gm m^{-2} day^{-1}. This is chiefly because such insecticides have been found to be highly toxic to earthworm (Wearly and Merriam, 1978). Similarly Heneghan et al., (1999) and Gonzalez and Seastedt (2001) found that excluding micro invertebrates slowed decomposition rate of leaf litter was slowed in humid tropical forests.

Relationship Between Abiotic Factors and Forest Floor Litter Decomposition and Turn Over

The dry matter dynamics of the forest floor litter biomass is strongly determined by rain fall, temperature and humidity of the forest. The abiotic environmental factors like rainfall, temperature and humidity had significant influence on litter production. Observations by Rajendraprasad *et al.* (2000) in the dry tropical sacred grooves of Kerala and Reich and Berchert (1984) in the tropical dry forest of Costaa Rica also have established strong positive influence of abiotic parametres on litter biomass. Facelli and Pickett (1991) reported that in the upland ecosystems, climatic conditions are the major determinant of the litter dynamics. The production and decomposition of dry matter in an ecosystem are also strongly related to light availability during the growing season (Jordan, 1971) and temperature and precipitation (Vogt *et al.*, 1986). The site-to-site variation in the litter production in the SBR can be explained as the effect of biotic and abiotic factors of the ecosystem, which in turn influence the chemical properties of the soil. This may also have selective influence of other biota on the litter (Xiong and Nilsson, 1997). The tropical environmental conditions such as soil and atmospheric temperature and humidity influence the moisture status of the system which in turn influences the litter dynamics.

Patterns of Litter Decomposition and Nutrient Release

Generally there are two steps in the litter decomposition, each one with different decomposition rates (Swift and Anderson, 1983). Berg and co-workers (Berg and Staaf, 1980; McClaugherty and Berg, 1987) have shown that in the initial stages (0 to 3 months) of leaf breakdown small soluble carbon molecules like starches and amino acids are lost first leaving behind the more recalcitrant molecules like lignin. Decomposition during this first phase is lipid because these molecules are easy to breakdown and energy rich. The second stage of decomposition – the reak down of lignin – is much slower because lignin consists of very large and complex molecules. This rapid initial breakdown followed by a longer period of slow decomposition results in a mass loss curve that resembles an exponential decay curve. During the process of decomposition, amount of different nutrients in the decomposing litter generally do not relate with litter biomass. For instances, in the tropical wet evergreen forest of Nelliampathy (Chandrashekara,1992) nitrogen and phosphorus showed much fluctuation during 1 year period, often exceeding the initial concentration potassium, calcium and magnesium showed a gradual decline with passage of a time the relative increase in the concentration of nitrogen and phosphorous in the leaf litter during the process of decomposition and their rapid fluctuation may be related to immobilization of these two nutrients by phylloplanemicroflora, so that they are released more slowly and at the same rate as organic matter loss (Gosz et al., 1973; Toky and Ramakrishnan, 1984). In contrast, labile element like potassium is released at a faster rate. It was also recorded that in general nutrient levels at the end of one year was in the order of N>P>Mg>Ca>K. However, when different species are considered mobility of the nutrients from decomposing litter may show different order. In this context, studies related to decomposition and nutrient release patterns in green leaf manure (consisting of leaves single species or a mixture species) will have significance as the information thus generated could be useful to understand the synchronization of nutrient release by green leaf manure and the nutrient up take by crop species.

Decomposition is a critical process for regulating nutrient cycling and production in all ecosystems. Though attempts are being made to assign economic value for some of the ecosystem services such as biomass production, aesthetic beauty of natural forests, pollination of crops and natural vegetation, etc., such an attempt has not been made in the case of litter decomposition and associated ecosystem services. In fact, litter decomposition is one among several other ecosystem services which are not traded in markets and hence do not have prices, and thus are interpreted as having no value when it comes to making decision about their use. In this context, quantitative evidence on economic value of litter decomposition needs to be generated. As prerequisite, methods for analysing the ecological economics of litter decompositions are need to be developed.

Forest floor litter biomass plays two principal roles in forest ecosystems; firstly it forms an inherent part of the nutrient cycle of the forest and is a major sourse of soil organic matter. Secondly presence of forest floor biomass acts as protection against microclimatic fluctuations. Low forest floor biomass leads to decline in fresh decaying organic matter in the soil surface layer and low nutrient status of soil in the transition and buffer areas of the reserve. As Similipal biosphere reserve is a northern tropical moist mixed deciduous forest and the tropical deciduous plants are badly adapted to low nutrient conditions of soil, litter addition treatments in the transition and buffer areas of the reserve may helpful in maintaining the soil nutrient status which can be achieved through increase in vegetation cover in the degraded conditions of the reserve.

Chapter-6

PHENOLOGY OF PLANT LIFE FORMS OF SIMILIPAL

Phenology is the study of the timing of recurring biological events, among phases of the plant species, which provide a background for collecting and synthesizing detailed quantitative information on rhythms of plant communities. Tropical plants with their high level of species diversity display phenological events such as leaf drop, leaf flushing, flowering and fruiting, etc. in relation to time and space (Justiniano and Fredericksen, 2000; Singh and Singh, 1992). Such periodic activities of plants in tropical environments have received much attention in recent years. Knowledge in the context of periodic activities of plants has helped to understand the influence of phenological events on feeding, movement patterns, and sociality of insects, birds and mammals (e.g. Foster 1982b; Leigh and Windsor, 1982). Recently, efforts have been made to know the importance of general community patterns in leafing, flowering and fruiting for many species, of which particular forest types are composed (Frankie, 1974a; Lieth, 1974; Opler, 1980). A considerable amount of information is available on the major phenological events of plant species from different parts of tropical America, Africa and south-east Asia including continental India have already available (Santapau, 1962; Malaisse, 1974; Monasterio and Sarmeinto, 1976; Liberman, 1982; Rai and Proctor 1986; Steven et al 1987; Bullock and Solis-Magallanes, 1990; Boojh and Ramakrishnan, 1982; Shukla and Ramakrishnan, 1982, 1984). However, phenodynamics and the variation in the distribution of phenological events, the phenology of tropical moist forest species of Odisha with respect to Similipal Biosphere Reserve (SBR) is only conducted by Mishra et al. (2006).

The displacement and adjustment of flowering and fruiting of different species in time and space for the efficient utilization of pollinators (Levin and Anderson, 1970) and seed dispersal agents (Janzen, 1970; Arroyo, 1979) are important for niche differentiation in a forest community. Recently, attention has also been paid to the relationship between phenological patterns and the strategies of vegetative growth and reproduction. An understanding of the phenological events in a forest community may reveal the structural organization of various types of resources in the ecosystem.

Sampling and Collection of Phonological Data

Permanent study plots were selected inside the biosphere reserve on the basis of aspect and level of anthropogenic disturbance. All groups of plant species were enumerated in each study plot. The plant specimens collected from the field were identified following the local floral guide or other floral guide as per example flora of Odisha (Saxena and Brahmam, 1994-96) and flora of Botany Bihar and Odisha (Haines ,1925). Phenological observations were made on species from all the sites. Individuals of each of the plant species were marked and tagged for each species. Whenever the required number of individuals was not available inside the permanent plot, additional individuals in adjacent area were also marked. Besides, observations on large populations by taking additional similar stands were also made to alleviate replication deficiency. If a given phenol-phase was observed in 5-10 per cent individuals of a species, it was considered to have initiated. For each tagged tree, records were made on leaf drop, leaf flushing, flowering and fruiting for over storey and under storey species. For ground flora records were made on vegetative growth, flowering, fruiting and seed maturation. All tree species were divided into two categories: (i) Over storey species consists of canopy and sub-canopy trees with a height of > 10m and (ii) under storey species with < 10m. Evergreen species continually produce at least small amount of new leaves throughout the year and do not show heavy leaf fall at a concentrated period whereas deciduous species has a marked leaf fall and leaf flushing at a concentrated period of the year.

Duration and Pattern of Activity

Both brief and extended activity was observed for the periodicity of leaf flushing, flowering and fruiting activity by individuals of a species in a population. Brief activity extends for 2 weeks or less while extended activity refers to periods more than 2 weeks. A more or less continuous flowering and fruiting activity throughout the year is referred to as continuous activity. The term seasonal and extended activity refers to flowering/fruiting occurs during a given period and extending into more than one period respectively. Marginal activity refers to species that have their activity occurring during transition period of seasonal changes. When some individuals of a tree species are in flowering/fruiting simultaneously is referred to as synchronous activity (S). The species showing flower/fruit development during a distinct period is known as asynchronous (A).

Fruit Maturation Activity

There are 2 categories of fruiting activity, i.e., rapid and lengthy. Rapid (r) can be characterized as fruit maturation periods of 4 months or less following fertilization and those more than 4 months is termed as lengthy (L).

Phenophases of Overstorey and Understorey Species of Similipal

In this chapter explanation on phenophases was made about 131 species of overstorey, understorey and ground flora/herbaceous species of Similipal. Among the 88 species of overstorey and understorey category 53 were deciduous (29 overstoreys and 24 understoreys), 16 were semi-evergreen (6 overstoreys and 10 understoreys) and 20 were evergreen species (16 overstoreys and 4 understoreys). The forest does not maintain its green appearance throughout the year because majority of the species are deciduous in this forest. However in the wet months of July to October deciduousness of the forest is not so conspicuous due to reduced leaf fall in comparison to the drier and cool months. The general phenological stages of some selected overstorey species, understorey species and ground flora are presented in Table -6.1, Table-6.2 and Table-6.3, respectively. Seasonal activities of the over and under storey species have been discussed under leaf drop, leaf flushing, flowering and fruiting activities. However, in case of ground flora or herbaceous species discussion was made on seed germination, vegetative growth, flowering, fruiting, seed germination and death.

Leaf Drop

Leaf drop may totally or partially depending upon the species. In some truly deciduous taxa, all or most of the old leaves got abscised before the arrival of new ones and the tree was bare for a period of weeks or to few months. Examples of this category are *Anogeissus latifolia, Adina cordifolia, Aegle marmelos, Albizia marginata, Bombax ceiba, Croton roxburghii, Diospyros melanoxylon,* etc. In other species such as *Ficus benghalensis, Albizia procera, Barringtonia acutangula, Diospyros sylvatica, Garuga pinnata, Ficus microcarpa, Mangifera indica* and *Mitragyna parviflora,* leaf fall and leaf flushing processes slightly overlapped in the same tree. In evergreens old leaves were abscised over a period of time throughout the year, thus, retaining a steady population of functional leaves all the time. Majority of the species start leaf shedding in dry months, i.e., from January and extending up to May and being low in other months. The peak of leaf drop was recorded in February to March in overstorey species and March to April in understorey species (Table-6.1 and 6.2). Arjunan and Pannammal (1993) stated that leaf drop is delayed due to rain and high temperature and advanced due to drought and low temperature.

Table 6.1: Phenology of overstorey species of Similipal biosphere reserve.

Name of the plant species	VT	Leaf drop	Leaf flushing	Flowering	Fruiting
Anogeissus latifolia	D	March	April	June-September	December-January
Adina cordifolia	D	February-March	April	June-July	February-May
Aegle marmelos	D	March- April	May	March-April	May-June
Albizia procera	SE	May-June	June	August-September	Dec,Jan, Feb,March, April, May
Albizia marginata	E	January-February	March	May-June	October-April
Albizia lebbeck	D	January	April	February-April	March-June
Alangium salvifolia	D	January-February	March	February–May	March-May
Anthocephalus cadamba	E	January-December/ April-March		August-October	November,December, January
Buchanania lanzan	D	December,January	January	February–March	March-April
Bridelia retusa	E	January-December,March-April		August-October	November, December, January
Bombax ceiba	D	February-March	April	March–April	March-June
Baringtonia acutangula	SE	February-March	April	May	September
Croton roxburghii	D	January-February	March	January-February	March-April
Careya arborea	D	February-March	March	March-May	July
Cassia fistula	D	March-April	April	May-August	January-December/ April*

Name of the plant species	VT	Leaf drop	Leaf flushing	Flowering	Fruiting
Dalbergia sisoo	D	December ,January , February, March	April	March-April	June –July
Diospyros melanoxylon	D	January- February	March	March-April	June-July
Dalbergia latifola	D	April-June	June	March-April	June-July
Dillenia pentagyna	D	February-May	May	March-April	May-June
Diospyros embryopteris	E	January-December/ March-April	April	May	March –June
Diospyros malabarica	E	January-December/ March-April	April	May	May
Diospyros sylvatica	SE	March-April	April	May	January-March
Ficus benghalensis	SE	March	May-June	May-August	April-June/December-February
Ficus hispida	SE	March	April-May	January-December	October-December
Ficus microcarpa	SE	March	March –April	—	October-April
Ficus religiosa	D	December-February	March-April	—	June-October
Garuga pinnata	SE	January-february	February-March	February - Mach	March-May
Gmelina arborea	D	February-March	March	February-March	May-June
Kydia calycina	D	March	March-May	September-November	December-May

Name of the plant species	VT	Leaf drop	Leaf flushing	Flowering	Fruiting
Lannea corromandelica	D	March-April	April	March-April	
Lagerostroemia parviflora	D	February-March	March	April-May	December-January
Madhuca indica	D	March-April	April	March-April	June-July
Mangifera indica	SE	April-May	June	January-March	May-June
Mitragyna praviflora	SE	April-June	April-June	May-June	March-April, November
Melia dubia	D	March-April	May	June	June
Miliusa velutina	D	April	May-June	May-June	June-July
Michelia champaca	SE	March-May	March-October	April-may	July
Pterocarpus marsupium	SE	April-May	May-June	October	December-January
Protium serratum	D	March-April	April	April	May-August
Phoebe wightii	SE	March	April	April-May	May-June
Phoebe lanceolata	SE	February-March	April	April-May	May-June
Shorea robusta	D	March	April	April-May	June-July
Syzygium cumini	SE	February-March	April	March-April	May-June
Sterospermum suaveolens	D	March	April	April-May	September-February
Schleichera oleosa	SE	March	April	March	June
Suregada angustifolia	E	January-December/ January-February*	February	March-April	—

Name of the plant species	VT	Leaf drop	Leaf flushing	Flowering	Fruiting
Samanea saman	E	January-December	—	March-April	May-July
Schrebera swietenoides	SE	February-March	April-May	April-May	May-June
Terminalia alata	D	March-May	June-July	May-June	July-October
Terminalia chebula	D	February-March	April	April-May	November-February

Source: Mishra et al. (2006)

Table 6.2: Phenology of understorey species of Similipal.

Name of the plant species	VT	Leaf drop	Leaf flushing	Flowering	Fruiting
Atrocarpus lacucha	SE	March	April	December,April	May,October-November
Alangium salvifolium	(D)	October-November	January-February	January,March, December	January-May
Bauhinia malabarica	(E)	January-February	March	March-June	May-November
Bauhinia parpurea	(SE)	April	June	July-August	October-November
Bauhinia variegata	E	March	April	February-March	April-May
Crateva religiosa	(D)	January-March	March	March-April	June
Casearia graveolens	(D)	May-June	June	May-June	May-July
Cartunaregum spinosa	(D)	March-April	May	May-June	September-December
Chionanthus intermedious	(D)	March-April	May	January-April	March-May

Name of the plant species	VT	Leaf drop	Leaf flushing	Flowering	Fruiting
Casearia elliptica	(D)	February-April	April	April;May-September	April-may
Cleistanthus collinus	D	March-April	April	April-May,September	March-April
Euonymus glaber	E	—	—	May	March-June
Ficus glomerata	(D)	October-November	December-January	April	March-June
Glochidion lanceolarium	E	—	March	March-May	September-June
Glochidion velutinum	(D)	March-April	April-May	April-May	June-August
Gardenia latifolia	(D)	March-April	May	April	December-June
Gardenia gummifera	(D)	April	May	March-May	June-August
Homalium nepalens	(D)	March-April	May	May-June	—
Hymenodictyon excelsum	(D)	November-May	May	August	January
Holarrhena antieysentrica	(D)	February-April	April	May-July	December
Hypecanthera stricta	E	January-December	—	April-August	November-May
Ixora parviflora	(E)	January-December	—	March-May	May-June
Ligustrum gambler	(SE)	—	—	June-August	December-January
Nyctanthes arbortristis	(D)	April-May	May	September-November	December- January
Ochna obtusata	(D)	February-March	April	March- May	June-August
Phyllanthus emblca	(D)	April	May	September	September

Name of the plant species	VT	Leaf drop	Leaf flushing	Flowering	Fruiting
Prunus ceylanica	(SE)	September-October	November	August	November-February
Securinega virosa	(D)	March-April	April	May-September	May-September
Symplocos cochinchinensis	(E)	January-December/ May-June*	March-February	January-March	April-June
Wendlandia tinctoria	(E)	January-December/ January –February*	March-April	September-October	March-June
Wendiandia exerta	(E)	January-December/ February-March	April-May	January-March	March-April
Ziziphus rugosa	(D)	January-March	May	March-April	April-May
Ziziphus mauritana	D	January-February	April	February- March	May-July
Terminalia bellirica	(D)	February-March	April-September	March-May	January-February
Trewia nudiflora	(D)	December,January-February	March	January-March	November-December
Terminalia arjuna	(D)	March	April	April-July	May-August
Vitex leucoxylon	(E)	—	—	May-June, October	February
Xylia xylocapra	(D)	April-may	May	April-may	February-april

Source: Mishra et al. (2006)

Table 6.3: Phenology of ground flora/herbaceous species of Similipal.

Name of the species	Seed Germination	Vegetative growth	Flowering	Fruiting	Seed Maturation	Death
Ammania baccifera	July	August	August, September	October, November	November	December
Amaranthus spinosus	June, July	August, September	August, September, October, November	October, November	November	December
Aneilema nudiflorum	July	July	—	October	October, November	November
Aeschynomene indica	July, August	August	August, September	October	October, November	November
Abutilon indicum	July	July, August September	September, October, November	October ,November	November, December	December
Ageratum conyzoides	July	August, September	September, October, November	October, November, December	November December	December
Alternanthera sessilis	July	August, September	August, September, October, November	October, November	October, November, December	December
Achyranthes aspera	July	August	August, September, November	September, November	November, December	December
Cassia tora	July	July, August	August, September	October	October, November	November, December

Name of the species	Seed Germination	Vegetative growth	Flowering	Fruiting	Seed Maturation	Death
Cyanotis axillaris		July, August	August, September	October, November	November	November, December
Commelina erecta	June , July	July, August	August, September	October, November	November	December
Croton bonplandianus	July	August, September	August, September, October, November	October, November	November, December	December
Cyperus flavidus	June	July, August, September	August, September	October	October, November	November, December
Cyperus iria	July	July, August	August ,September	September, October	October, November	November
Cyperus distans	July	August	September	October	October, November, December	—
Cyperus haspen	June ,July	August	September	—	October, November	November
Cleome viscosa	July	July, August, September	September	—	October, November	November
Corchorus aestuans	June ,July	July, August	August, September	October	October	November
Centella asiatica	July	July, August	September	October, November	October, November	November, December
Celosia argentiea	—	August	September	October, November	October, November	December

Name of the species	Seed Germination	Vegetative growth	Flowering	Fruiting	Seed Maturation	Death
Cynodon dactylon	July	August	August, September	October	October, November	November, December
Dactyloctenium aegypticum	June ,July	July, August	August, September	October	October	November
Desmodium triflorum	June ,July	August	August, September, October, November	October, November	November	December
Echinochloa colonum	July	July, August	September	October	October, November	November, December
Echinochloa colona	July	August, September	September	October	October, November	November, December
Eclipta alba	July	August, September	September, October, November	October, December, November	November, December, January	January
Eriocaulon quinquangulare	July September	September, October	October, November	November, December	December, January	January
Euphoribia hirta	June ,July	August, September	September, October, November	October, November, December	November, December	December, January
Euphorbia nivulia	June ,July	August	September, October, November	October, November, December	November, December	December, January
Fimbristylis dichotoma	June ,July	August	August, September, October	October, November	November	November, December

Name of the species	Seed Germination	Vegetative growth	Flowering	Fruiting	Seed Maturation	Death
Indigofera linifolia	July	July, August	August, September	October, November	October, November	November, December
Ischaemum rugosum	July	August, September	September	November	October, November	November, December
Kyllinga cylindrica	June ,July	July	August, September	—	October, November	November, December
Ludwigia hyssopifolia	July, August	August, September	October, November	—	December, January	January
Paspalum scrobculatum	July	August, September	September, October	October	October, November	December
Phyllanthus fraternus	July	August, September	September	—	October, November	November, December
Rungia pectinata	August	September, October	October, November	December	December, January	January
Sida veronicaefolia	July	August	August, September	October	October, November, December	November, December
Sida acuta	July	August	September	October	October, November	November, December
Scirpus grosssus	June ,July, August	August, September	August, September	—	October, November	November, December
Xanthium strumarium	June ,July, August	July, August, September	July, August, September	November	October, November, December	November, December, January

(Source: Present investigation, 2013)

Leaf Flushing

Trees of tropical dry deciduous forest of SBR, responded to leafing during dry season (Mishra et al., 2006). Moisture appears to be a major determinant of the timing of leaf flush in dry tropical forest (Lieberman, 1982). The patterns of leaf formation were strongly influenced by rainfall, as expected for tropical semi-arid vegetation Bullock and Solis-Magallanes, 1990). Most deciduous species of dry monsoon forests in India form new leaves one to two months before the first monsoon rains, during the hottest and driest part of the year around the spring equinox (Elliot et al., 2006). Dry season flushing was also more in Similipal as compared to other seasonal forest covers of India (Bhooj and Ramkrishnan, 1981; Murali and Sukumar, 1994). The present observation also show that in deciduous species of both overstorey and understorey category, leaf flush or leaf initiation became more pronounced from March (18 species with 20%) and peaks in April and May (60 species with 67%) in the dry season (Table-6.1 and 6.2). In evergreen species of SBR, most of the leafing activities happen during pre summer to summer season as rainfall plays a significant negative influence. In this biosphere reserve, canopy trees leaf initiation begins from December and with a peak in February or March. This kind of response is due to adaptive nature of species and similar observations from other studies (Murali and Sukumar, 1994; Singh and Singh, 1992; Singh and Khushwaha, 2005; Kikim and Yadava, 2001) have been reported. At the community level, certain species have evolved completely different approaches to leaf flushing which may be due to intrinsic factors (Boojh and Ramakrishnan, 1981).

Flowering Activity

Flower initiation differs with respect to vegetation type. In evergreen and semi-evergreen group both in the overstorey and understorey category flower initiation begins during end parts of monsoon or early parts of winter and attains peak during post-winter to pre-summer (Table-6.1 and 6.2). However, in deciduous group most of the species in overstorey and understorey category initiate flowering in leafless period and few species simultaneously leafing and flowering in summer period. As a whole irrespective of vegetation type it is also clear from the figure that in most of the species flower initiation begins in January and continues till June with peaks during April to June. As evergreen trees initiated flowering in the winter period is similar to other studies (Saha, 2007). Whereas deciduous trees initiate flowering during the beginning of the dry season at the time when most of the trees were leafless or leaf flushing stage because flower initiation can advertise to pollinators as they get pollinated as seen in other tropical forest (Murali and Sukumar, 1994;). The synchronization of flowering with leaf flushing seems to be related to moisture, temperature and photoperiod (Bhooj and Ramkrishnan, 1981; Murali and Sukumar, 1994). Cool and dry winter period is responsible for maximum leaf drop whereas increase in temperature during warm and dry periods induces the leaf flushing and flowering in most of the species.

Fruiting Activity

Fruiting activity of both overstory and understorey species of the reserve was mostly concentrates around May and June than in December (Table-6.1 and 6.2). The peak period of fruit maturity study was observed during winter and summer. The fruit development period for different species of both layers varies from 4 to 28 weeks. A majority of the species in both the categories showed rapid fruiting activity. Next to rapid fruiting activity a larger proportion of species recorded lengthy fruiting behavior but only very few species had multiple fruiting behaviour (Table-6.1 and 6.2). Almost all tree species had phenological patterns that synchronized flowering and fruiting in the dry months, i.e., April, May and June. Most of the species flowered at the beginning of April and fruited near the end of May and beginning of June, needing only a short time for the development of fruits. Rest of the species flowered during April and May, fruited during December, with a moderate amount of time required for fruit development. Flowering and fruiting at hottest summer, i.e., April and May have selective advantage. It is more efficient to transfer assimilates directly into growing organs rather than having to store them and mobilize and translocate them latter (Wright and Schaik, 1994). This perhaps establishes that increased temperature favours formation of fruits in most of the overstorey and understorey species.

In both the category of woody tree species ripening of fruits began in the later part of rainy season and continued up to end of cool and dry winter period. This is due to the difference in time taken for fruit maturation. Forty-seven out of 57 overstorey species showed extended and rapid fruiting activity, while only 7 and 3 species showed extended lengthy and extended rapid and multiple fruiting activity, respectively. Similarly out of 33 understorey species 27, 5 and one species showed extended rapid, extended lengthy and, extended and multiple fruiting activity, respectively. Similar observation has also been reported for tropical forests (Bullock and Solis Margallenus, 1990; Frankie et al., 1974) of Himalaya. It has also been reported that minimal pest pressure (Aide, 1988) and maximal activity of pollinating insects (Foster, 1996) may occur during dry season. Also fruit production at the end of dry season ensures that seedlings are not immediately exposed to water stress.

Ranges of Interphenophase Durations

Interphenophase durations between different phenological events of overstorey and understorey tree species belongs to different category viz. evergreen, semi evergreen and deciduous species of Similipal are vary from each other. The shortest range was for maturation of leaves and the longest for fruiting. Retention of flowers in evergreen species was for a longer period than that in deciduous species. As leaf drop and initiation is concerned in SBR, the dry deciduous species responded well to leafing during dry season, i.e., in summer (Mishra et al., 2006). This may be attributed to the triggering effect of the rising temperature (Walter, 1968) and increase in length of photoperiods (Lawton and Akpan 1968; Njoku 1964). Dry season flushing was more in most of the species, which is similar to earlier studies (Frankie et al., 1974; Van Schaik, 1986; Murali and Sukumar, 1994; Khushwaha and Singh, 2005). Bolivian dry forest canopy species lost their leaves during the dry season especially from January to February and to some extent in March may be due to direct influence of the decline in soil moisture and increasing water stress conditions (Justiniano and Fredericksen, 2000). Several studies suggest low temperatures and short days are the main factors involved in triggering leaf senescence in deciduous trees.

The results are also in conformity with Singh and Singh (1992) in which the initiation of leaf fall coincides with the onset of the post-monsoon low temperature dry period and can be a mechanism maintaining turgidity of shoots. However, in evergreeen tree species of SBR, most of the leafing activities happen during pre summer to summer season as rainfall plays a significant negative influence. In attains, canopy trees leaf initiation begins from November or December and a peak in January or February. This kind of response is due to adaptive nature of species and similar observations from other studies (Murali and Sukumar, 1994; Singh and Singh, 1992; Singh and Khushwaha, 2005; Kikim and Yadava, 2001) have been reported. At the community level, certain species have evolved completely different approaches to leaf flushing and leaf fall which may be due to intrinsic factors (Boojh and Ramakrishnan, 1981).

Phenophases of Ground Flora

Seasonal ecological phenomenon has been described by superimposing phonological calander over civic calander. Observations on the different phases of life cycle of a plant like seed germination, vegetative growth, flowering, fruiting and seed maturation in their temporal occurrence are used to prepare a phonological calendar. These phonological observations were recorded in the field through several visits at regular intervals and presented as phenograms. These provide a good information about the ecological life and behavior of an organism. The phonological behavior of ground flora was found to be closely related with the climatic cycle. Most of the ground flora or herbaceous species show their vegetative and reproductive growth during rainy season, a few in winter and virtually none in summer. The herbaceous species of the reserve mostly emerge in the month of July after rains. Flowering starts in most of the species from first week of September and seed shedding from September to October. Most of the species are rainy season annuals show their vegetative growth during August and September. During September and October most of the plants are studded with flowers and show their fruiting in late October. With the onset of winter most of the rainy season ground flora/herbaceous species start declining and by November most of them dry up. The ground flora/herbaceous species of winter season start germinating during late rains and flower profusely in the middle or at the end of winter season. The constant wettings of seed coat during the rainy period followed by the sudden rise in diurnal fluctuations of temperature appear favourable for germination of seeds of winter annuals. Thus the ground flora or herbaceous species of Similipal represent two distinct seasonal aspects, i.e., rainy and summer aspect. It is thus evident that temperature, humidity and precipitation are potent climatic factors that control the growth of plants to a great extent. Moreover the change in physical environmental factor on the forest floor of the biosphere reserve may affect the diversity of ground flora or the herbaceous species (Table-6.3).

Future Activities on Phenophases of Plant Life Forms

In Similipal, high diversity of overstorey and understorey tree species and multiple triggering factors for bud break (vegetative and flower) are major causes for weak synchronization of community-level phenological events with the onset of growth-suppressive seasonal drought or growth-promoting rainy season. Long-term quantitative documentation of phenological patterns in a tree species is required in a standardized way to assess the range of variation within and between species, and to segregate them as per functional types. Study on major objectives like (a) to screen-out tree species into functional types based on deciduousness and triggering factors; (b) to evaluate the interrelationships between vegetative and reproductive phenological events; (c) to explore the cause of vegetative and flower bud break in different species; (d) to model how different functional types interact with seasonal variation in climatic factors and soil moisture storage through their water relations and phenological events; (e) to examine functional wood anatomy and root systems in different functional types; (f) to evaluate the patterns of resource use and to establish the links among water relations, phenology and primary productivity will be emerge as important focus points for future ecological research, not only because of its relevance to forest structure and functions, but more importantly due to its potential to address critical questions in modelling, monitoring, and climate change.

Chapter-7

HERPETOFAUNAL DIVERSITY OF SIMILIPAL

Biotic surveys and information on geographic distribution of the species are essential for Biogeographic studies, and can provide a baseline for ecological and biodiversity conservation as whole. This information is particularly important for amphibians and reptiles, as they are among the most threatened vertebrate animal in the world. Deforestation and fragmentation are among are among the most important human-induced threatening factors for amphibian and reptiles, and monitoring habitats under degradation is urgent to assess real threats on species of these groups. Accordingly, the identification of amphibians and reptiles and the study of their ecological characteristics are decisive for the success of actions directed to biodiversity conservation. Amphibians play a very important role in the food chain of both terrestrial and aquatic ecosystems. The general ecological importance of amphibians lies in them being predators acting as primary and secondary carnivores on insects, some which crop pests or disease vectors (Behangana, 2004). Amphibians are widely considered to be useful as indicator species (Sheridan and Oloson, 2003). A considerable amount of data related to richness and composition of communities may be assembled through appropriate bibliography and field inventories. Herpetological inventories can offer a wider vision of distributional patterns of a large number of species which optimizes the comprehension efforts of species distribution related to different environmental variables.

The herpetofauna of Odisha is represented by more than 155 species, including 26 species of frogs and toads, 3 species of crocodilians, 16 species of turtles, 30 species of lizards and 80 species of snakes (Dutta et al. 2009). Similipal Biosphere Reserve is one of the richest tropical forest and species list updates and accurate information on the geographic distribution of species are necessary for baseline studies in ecology and conservation of this area. Here, we present an updated list of the diversity of amphibians and reptiles of Similipal Biosphere Reserve, and actualized information on their distribution and conservation status. Although some studies have discussed the amphibian and reptiles of Similipal Biosphere Reserve, most herpetological list came from Dutta et al. (2009), reported 21 species of frogs, and 60 species of reptiles, comprising 1 species of crocodile, 6 species of turtles, 20 species of lizards and 33 species of snakes from Similipal Biosphere Reserve.

The main purpose of this study was to survey the reptiles and amphibians of Similipal Biosphere Reserve, Odisha. This study is vital for the amphibian and reptile biodiversity inquiry in Similipal Biosphere Reserve because it provides information that could be used for future herpetofauna conservation programs. If no effort is made in order to enhance herpetofauna conservation, we may lose a few species before we can thoroughly study them and we can even lose some species before we actually discover them. The species list is based on species observed during the different Faunal Monitoring project in Similipal Biosphere Reserve.

Herpetological Inventories

Field surveys were conducted during the day and sampling activities were carried out 2 to 3 h during night-time by field parties comprising 2-3 persons. These activities involved walking along a 1.5 km length of the riparian zone of different rivers and swampy areas that flows through the area and along 3 km of forest trails criss-crossing the forest area, looking for any reptiles or amphibians. Reptiles were searched in the respective microhabitats during day and night-time. Active search along the road side, below rock boulders, below rotten logs, in leaf litters, near water bodies and also on bushes yielded good results. Inventories were done in all available landscape element in the region, including areas of different forest patches, pastures, agricultural field, rivers and streams and human settlements. We also included road kill records and those of species found during field work of other research projects carried out in the region.

Herpetofaunal Diversity of Similipal Biosphere Reserve

Twenty species of amphibians and 60 species of reptiles were recorded in total from Similipal biosphere reserve (Table-7.1 and 7.2). The families Colubridae (22 species) was diverse in the reptiles assemblage, while the families Dicroglossidae (8 species) and Microhylidae (5 species) were the most diverse families for amphibians (Table-7.1 and 7.2).

Lizards were represented by 19 species belonging to seven different families. Family Scincidae was represented by six species and had the highest species number followed by families Gekkonidae represented five species, Agamidae comprising three species, Varanidae with two species and, Chamaeleonidae, Eublepharidae and Lacertidae comprised one species each.

The number of amphibian and reptile species found in the Similipal Biosphere Reserve is particularly high and one of the most species rich topical forests in terms of herpetofauna.

Conservation Measures

Amphibians and reptiles face other important threats in this region, directly related to human activities such as hunting, or killing out of fear. There are three edible species of frogs found in Odisha and two of them (*H. crassus* and *H. tigerinus*) are found in Similipal Biosphere Reserve. However, there is no commercial exploitation of frogs in Odisha and both the above species are included in Appendix-II of the Convention of International Trade in Endangered Species of Wild Fauna and Flora (CITES). In addition, the export of these species requires permits under CITES and the Indian Wildlife (Protection) Act, 1972. The above regulations prevent the exploitation of frogs, but this does not mean that the frog populations are safe in the state. Most of the species are seasonal breeders and they migrate to temporary rainwater pools for breeding. Due to utilization of virgin lands for human settlement, a vast majority of these agricultural and open fields are gradually vanishing. Hence, there is a threat to future survival of some frog species.

Reptilian conservation in Odisha is confined to all the three species of crocodilians, sea turtles, a few species of snakes (Python and King Cobra) and lizards (Varanus). Since 1976, the state Forest Department has formed a separate Wildlife Wing which is involved in protection, management and conservation of wild animals in all the protected areas. Similipal Biosphere Reserve Biosphere Reserve provides protection to some mega species of reptiles (Mugger Crocodile, Python, King Cobra and Land Tortoise). However, most other small species of reptiles are also protected indirectly in the Biosphere reserve. Though few snakes (Common Cobra, King Cobra and Rat Snake), lizards (Varanus and Chamaeleon) and turtles (soft-shell) are captured by tribals for various uses (non-commercial utilization), their population status in Similipal Biosphere Reserve is yet to be assessed. The most effective way of conserving these animals in the locality is through creation of awareness among the local inhabitants.

Table 7.1: Updated list of the amphibians in Similipal biosphere reserve

SL. No.	Common Name	Scientific Name
	Family: Bufonidae	
1	Common Asian Toad	*Duttaphrynus melanostictus* (Schneider, 1799)
2	Ferguson's Toad	*Duttaphrynus scaber* (Schneider, 1799)
3	Marbled Toad	*Duttaphrynus stomaticus* Lutken, 1864
	Family: Dicroglossidae	
4	Indian Skipper Frog	*Euphlyctis cyanophlyctis* (Schneider, 1799)
5	Syhadra Cricket Frog	*Fejervarya syhadrensis* (Annandale, 1919)
6	Dutta's Cricket Frog	*Fejervarya odishaensis* (Dutta, 1997)
7	Indian Bull Frog	*Hoplobatrachus tigerinus* (Daudin, 1802)
8	Jerdon's Bull Frog	*Hoplobatrachus crassus* (Jerdon, 1853)
9	Short-headed Burrowing Frog	*Sphaerotheca breviceps* (Schneider, 1799)
10	Indian Burrowing Frog	*Sphaerotheca rolandae* (Dubois, 1983)
11	Dobson's Burrowing Frog	*Sphaerotheca dobsonii* (Boulenger, 1882)
	Family: Ranidae	
12	Fungoid Frog	*Hylarana malabarica* (Tschudi, 1838)
	Family: Microhylidae	
13	Painted Balloon Frog	*Kaloula taprobanica* (Parker, 1934)
14	Ornate Narrow mouthed Frog	*Microhyla ornata* (Dumeril &Bibron, 1841)
15	Variable Ramanella	*Ramanella variegate* (Stoliczka, 1872)
16	Grey Ballon Frog	*Uperodon globulosus* (Gunther, 1864)
17	Marbled Ballon Frog	*Uperodon systoma* (Schneider, 1799
	Family: Racophoridae	
18	Common Indian Tree frog	*Polypedates maculatus* (Gray, 1834)
19	Dubois's Tree Frog	*Polypedates teraiensis* (Dubois, 1987)
20	Similipal Biosphere Reserve Bush Frog	*Philautus similipelensis* Dutta, 2003

Table: 7.2: Updated list of the Reptiles in Similipal Biosphere Reserve

SL. No.	Common Name	Scientific Name
	Order: Testudines	
	Family: Bataguridae	
1	Peninsular Tent Turtle	*Pangshura tentoria* (Gray, 1834)
2	Indian Roofed Turtle	*Batagur tecta* (Gray, 1831)
3	Tricarinate Hill Turtle	*Melanochelys triccarinata* (Blyth, 1856)
4	Eastern Black Turtle	*Melanochelys trijuga indopeninsularis* (Annandale, 1913)
	Family: Testudinidae	
5	Elongated Tortoise	*Indotestudo elongata* (Blyth, 1853)
	Family: Trionychidae	
6	Indian Flapshell Turtle	*Lissemys punctata* (Bonnaterre, 1789)
	Order: Crocodylia	
	Family: Crocodylidae	
7	Mugger Crocodile	*Crocodylus palustris* Lesson, 1831
	Order: Sauria	
	Family: Agamidae	
8	Indian Garden Lizard	*Calotes versicolor* (Daudin, 1802)
9	Indian Rock Lizard	*Psammophilus blanfordanus* (Stoliczka, 1870)
10	Fan-throated Lizard	*Sitana ponticeriana* Cuvier, 1844
	Family: Chamaeleonidae	
11	Indian Chamaeleon	*Chamaeleo zeylanicus* Laurenti, 1768
	Family: Eublepharidae	
12	East Indian leopard gecko	*Eublepharis hardwickii* Gray, 1827
	Family: Gekkonidae	
13	Clouded Ground Gecko	*Cyrtodactylus nebulosus* Beddome, 1870
14	Spotted Indian house gecko	*Hemidactylus brookii* Gray 1845
15	Indian house gecko	*Hemidactylus flaviviridis* Ruppel, 1840
16	Bark gecko	*Hemidactylus leschenaultia* Duméril & Bibron 1836
17	Smooth house gecko	*Hemidactylus frenatus* Schlegel, 1836
	Family: Lacertidae	

SL. No.	Common Name	Scientific Name
18	Snake-Eyed Lacerta	*Ophisops jerdonii* Blyth, 1853
	Family: Scincidae	
19	White-Spotted Supple Skink	*Riopa albopunctata* (Gray, 1846)
20	Common Snake Skink	*Lygosoma punctata* (Gmelin, 1799)
21	Common Indian Skink	*Eutropis carinata* (Schneider, 1801)
22	Eastern Bronze Skink	*Eutropis macularia* (Blyth, 1853)
23	Beddome's Grass Skink	*Eutropis beddomii* (Jerdon, 1870)
24	Limbless Skink	*Sepsophis punctatus* Beddome, 1846
	Family: Varanidae	
25	Common Indian Monitor Lizard	*Varanus bengalensis* (Daudin, 1802)
26	Yellow Monitor Lizard	*Varanus flavescens* (Hardwicke & Gray, 1827)
	Order: Squamata	
	Family: Boidae	
27	Indian Rock Python	*Python molurus molurus* (Linnaeus, 1758)
28	Common Sand Boa	*Gongylophis conicus* (Schneider, 1801)
29	John's Sand Boa	*Eryx johnii* (Russell, 1801)
	Family: Colubridae	
30	Common Vine Snake	*Ahaetulla nasuta* (Lacepede, 1789)
31	Buffstriped Keelback	*Amphiesma stolatum* (Linnaeus, 1758)
32	Red-necked Keelback	*Rhabdophis subminiatus* (Schlegel, 1837)
33	Banded Racer	*Argyrogena fasciolata* (Shaw, 1802)
34	Olive Keelback Water Snake	*Atretium schistosum* (Daudin, 1803)
35	Common Indian Cat Snake	*Boiga trigonata* (Schneider, 1802)
36	Forsten's Cat Snake	*Boiga foresteni* (Dumeril, Bibron & Dumeril, 1854)
37	Common Indian Trinket Snake	*Coelognathus helena* (Daudin, 1803)
38	Copper-Headed Trinket Snake	*Coelognathus radiatus* (Schlegel, 1837)
39	Common Indian Bronze-back	*Dendrelaphis tristis* (Daudin, 1803)
40	Smooth Water Snake	*Enhydris enhydris* (Schneider, 1799)

SL. No.	Common Name	Scientific Name
41	Ornate Flying Snake	*Chrysopelea ornate* (Shaw, 1802)
42	Common Wolf Snake	*Lycodon aulicus* (Linnaeus, 1758)
43	Barred Wolf Snake	*Lycodon striatus* (Shaw, 1802)
44	Twin-spotted Wolf Snake	*Lycodon jara* (Shaw, 1802)
45	Indian Green Keelback	*Macropisthodon plumbicolor* (Cantor, 1839)
46	Common Kukri Snake	*Oligodon arnensis* Shaw, 1802
47	Mock Viper	*Psammodynastes pulverulentus* (Boie, 1827)
48	Common Indian Rat Snake	*Ptyas mucosus* (Linnaeus, 1758)
49	Cantor's Black-headed Snake	*Sibynophis Sagittarius* (Cantor, 1839)
50	Checkered Keelback Water Snake	*Xenochrophis piscator* (Schneider, 1799)
	Family: Elapidae	
51	Common Indian Krait	*Bungarus caeruleus* (Schneider, 1801)
52	Baned Krait	*Bungarus fasciatus* (Schneider, 1801)
53	Monocellate Cobra	*Naja kaouthia* Lesson, 1831
54	Binocellate Cobra	*Naja naja* (Linnaeus, 1758)
55	King Cobra	*Ophiophagus hannah* (Cantor, 1836)
	Family: Typhlopidae	
56	Common Blind Snake	*Ramphotyphlops braminus* (Daudin, 1803)
57	Beaked Worm Snake	*Grypotyphlops acutus* (Dumeril, Bibron & Dumeril, 1844)
	Family: Viperidae	
58	Russell's Viper	*Daboia russelii* (Shaw & Nodder, 1797)
59	Saw-scaled Viper	*Echis carinatus* (Schneider, 1801)
60	Bamboo Pit Viper	*Trimeresurus gramineus* (Shaw, 1802)

Chapter-8

STATUS OF BIRDS IN SIMILIPAL

The birds have always fascinated man for their exquisite colouration and play an important role in forest ecosystem as potential pollinators, seed dispersers and scavengers. They are regarded as a viable indicator for biological biodiversity and changes in environmental conditions, for which their contribution to properly functioning ecosystems cannot be underestimated. The importance of local landscapes for avian conservation can only be understood by knowing the structure of the bird community in the region. Valuable information on factor influencing population dynamics, interactions, community structure, and conservation can be gathered by monitoring seasonal changes of avifauna.

There are about 9,026 species belonging to more than 1,800 genera, 182 families and 30 orders of the birds throughout the world while India harbours 1,166 under 405 genera. About 1,300 species of breeding, staging and wintering birds belonging to 88 families and 22 orders are reported from India by Manakadan and Pittie (2002). Around 500 species of birds are known from different parts of Odisha (Mishra et. al., 1996).

Ornithological History of Similipal

Earlier studies on the birds of Similipal were initiated by Ripley (1978) and Dev (1986). Dev (1986) listed 223 species recorded in Similipal. Manoj V. Nair recorded a total of 304 species from Similipal during his period as Deputy Director in Similipal Tiger Reserve. The recent update checklist of birds of Similipal was 361 species (Nayak and Naik 2014).The present study reported a total of 363 species of birds from Similipal.

Methods of Study of Avifauna

Observations were made, usually for a full day during the study period. Regular surveys were done by walking on fixed routs throughout the study area. Observations were made in the morning between 07:00 and 10:00 hr and/or in the afternoon 15:00 and 18:00 hr, depending on the light condition. Night observations were also conducted to listen and identified the nocturnal species like owls and nightjars. Recordings were not made at the time of heavy rains. Surveys were conducted twice a week. Birds were observed using 7 x 50 and 7 x 42 binoculars and identified. In different reservoirs in Similipal, different points were selected from which the whole sites could be observed for water birds. At each site birds were counted using a binocular before moving to the next point as rapidly as possible without disturbing the birds. Whenever necessary, surveys were conducted using non-mechanised country boats. During winters, all the major habitats were visited twice a month for monitoring wintering waterfowls. In the reservoirs total counts were carried out during the bird surveys. We have also included information from published literature, as cited in the text.

The status (movement and seasonality of occurrence), frequency of bird sightings in various habitats have been worked out basing on different parametres listed below.

Status: The status of birds were categorized by resident (R)-m found in all suitable habitats throughout the year, migrant (M)- found only during a specific season, winter migrant (W)-m found only winter season and summer migrant (S)-m found only summer season.

Frequency: Abundance was categorized by as common (C), uncommon (U), rare (R) and occasional (O).

Food habit: According to the feeding habits, birds were divided as piscivores (P) frugivores (F), omnivores (O), carnivores (C), grainivores (G) and insectivores (I).

Habitats: Different habitats where observations were carried include, aquatic habitat (A), aerial (Ar), open field (O), woodland (W), rocky outcrops (R) and scrub forest (S).

Composition of Birds of Similipal

A total of 363 species including 80 species of water birds, belonging to 67 families, were recorded (Table 8.1).Detail status, scientific and common names and description of the birds were given below, which shows that the Similipal supports a high diversity of birds. The Accipitridae family shows the highest species richness (34 species), followed by Sylviinae (18 species), Anatidae(16 species), Columbidae (15 species), Picidae (13 species), Ardeidae (12 species) and Strigidae (11 species). Of the species recorded in the study area, 300 species were resident and the remaining

63 species were migratory (Table-8.1). On the basis of abundance 84species can be considered as common, 170species are uncommon, 109species are rare. According food habit, 55 species can be considered as carnivore, 23 species are frugivore, 30 species were grainivore, 198 species are insectivore and 73 species were piscivores. A distinct seasonal variation in avian species richness was observed with a peak during the winter season.

Our study observed that Similipal supports a high diversity of birds. Most of observed species are breeding residents mainly due to occurrence of various types of microhabitat inside the Similipal, nearby different rivers and reservoirs. Due to abundance of rare and threatened species Similipal is very important for bird conservation in this part of the region. Woodlands are by far the most species rich habitat in Similipal due to the fact that this habitat type presents a large spectrum of available niches. It is important to note that woodland birds are more likely to escape detection than those occurring in rocky outcrops and open fields, such that the status of these woodland species should be viewed cautiously. Our goal here was to present a preliminary species list and assess the bird community of Similipal, but more effort to concentrate exclusively in this habitat type is necessary to achieve a more complete understanding of its composition and gain some sense of its meta-population dynamics.

Globally Threatened Species

Eight species are considered globally near threatened, five species are globally vulnerable, one species is endangered and three species are critically endangered (Table 8.2).Long-billed Vulture, Indian White-backed Vulture and Red-headed vultureare globally and nationally Critically Endangered;and Baer's Pochard, Pallas's Fish-Eagle, Greater Spotted Eagle, Pale-capped Pigeon and Bristled Grass-Warblerare globally vulnerable.

Water Birds and Winter Migrants

The presence of reservoirs in the study area creates a reasonable habitat for waterbirds and winter migrants. This addition of waterbirds to the overall numbers favourably contributed to the bird diversification in the area. Migratory waterbirds such as Great Crested Grebe *Podiceps cristatus*, Red-crested Pochard *Rhodonessa rufina*, Common Pochard *Aythya ferina*, Tufted Pochard *Aythya fuligula* and Common Snipe *Gallinago gallinago* were found to be using the reservoir basin areas for foraging. Some resident waterbirds like cormorants, egrets and herons inhabit in the reservoirs area. These birds are generally feed upon planktons, fishes, amphibians and invertebrates from reservoir. Birds such as Small Blue Kingfisher *Alcedo atthis*, White-breasted Kingfisher *Halcyon smyrnensis*, Small Bee-eater *Merops orientalis*, Common Swallow *Hirundo rustica*, Red-rumped Shallow *Hirundo daurica*, Paddyfield Pipit *Anthus rufulus* and Wagtail species were generalists present in all the habitats (Table-8.3).

Conservation Issues and Implications

Similipal's wetlands are the famous tourist spot and increase the tourist activity during months of December to February is threats to the birds of these areas. Cattle grazing, forest fire and use of forest wood as source of fuel by local people are also creating adverse conditions for the birds of the region. Grassland birds were particular conservation concern as their habitat is being rapidly and irreversibly converted to land for agriculture and cattle pasture. Many birds like peafowl, jungle fowl, hornbills and parrots are either trapped or poisoned by locals for consumption or cage birds.

Conservation of Similipal birds is not only of local importance but also global interest. Therefore various measures should be taken for the conservation of birds of the Similipal. Cattle grazing should be allowed in the controlled manner. Alternative fuel source should be made available to the local communities. Establishment of ecotourism committees with the help of local people and conducting awareness programs by the forest departments on a regular basis would be an effective step in the avian diversity conservation in Similipal.

Table-8.1: Abundance, status, food habit and habitat of avifaunal species of Similipal.

Family and Common Name		Scientific Name	Abundance	Status	Food habit	Habitat
Family: Podicipedidae						
1.	Little Grebe	*Tachybaptus ruficollis* (Pallas, 1764)	C	R	P	A
2.	Great Crested Grebe	*Podiceps cristatus* (Linnaeus, 1758)	R	M, W	P	A
3.	Red-necked Grebe	*Podiceps griseigena* (Boddaert, 1783)	R	M, W	P	A
Family: Phalacrocoracidae						
4.	Little Cormorant	*Phalacrocorax niger* (Vieillot, 1817)	C	R	P	A
5.	Indian Shag	*Phalacrocorax fuscicollis* Stephens, 1826	UC	R	P	A
6.	Great Cormorant	*Phalacrocorax carbo* (Linnaeus, 1758)	R	M	P	A
Family: Anhingidae						
7.	Darter	*Anhinga melanogaster* Pennant, 1769	R	R	P	A
Family: Ardeidae						
8.	Little Egret	*Egretta garzetta* (Linnaeus, 1766)	C	R	P	A
9.	Large Egret	*Casmerodius albus* (Linnaeus, 1758)	UC	R	P	A
10.	Median Egret	*Mesophoyx intermedia* (Wagler, 1829)	UC	R	P	A
11.	Large Egret	*Casmerodius albus* (Linnaeus, 1758)	UC	R	P	A
12.	Cattle Egret	*Bubulcus ibis* (Linnaeus, 1758)	C	R	P	A, O
13.	Indian Pond-Heron	*Ardeola grayii* (Sykes, 1832)	C	R	P	A
14.	Black-crowned Night-Heron	*Nycticorax nycticorax* (Linnaeus, 1758)	R	R	P	A
15.	Little Green Heron	*Butorides striatus* (Linnaeus, 1758)	UC	R	P	A
16.	Purple Heron	*Ardea purpurea* Linnaeus, 1766	R	R	P	A
17.	Grey Heron	*Ardea cinerea* Linnaeus, 1758	UC	R	P	A
18.	Black Bittern	*Dupetor flavicollis* (Latham, 1790)	R	R	P	A

Family and Common Name	Scientific Name	Abundance	Status	Food habit	Habitat
19. Yellow Bittern	*Ixobrychus sinensis* (Gmelin, 1789)	R	R	P	A
Family: Ciconiidae					
20. Asian Openbill-Stork	*Anastomus oscitans* (Boddaert, 1783)	C	R	P	A, O
21. Painted Stork	*Mycteria leucocephala* (Pennant, 1769)	R	M, W	P	A
22. White-necked Stork	*Ciconia episcopus* (Boddaert, 1783)	R	M, W	P	A
Family: Threskiornithidae					
23. Black Ibis	*Pseudibis papillosa* (Temminck, 1824)	UC	R	P, I	A, O
24. Oriental White Ibis	*Threskiornis melanocephalus* (Latham, 1790)	R	M, W	P, I	A, O
Family: Anatidae					
25. Lesser Whistling-Duck	*Dendrocygna javanica* (Hors eld, 1821)	C	R	P	A
26. Large Whistling-Duck	*Dendrocygna bicolor* (Vieillot, 1816)	UC	M, W	P	A
27. Spot-billed Duck	*Anas poecilorhyncha* J.R. Forester, 1781	UC	M, W	P	A
28. Red-crested Pochard	*Rhodonessa rufina* (Pallas, 1773)	C	M, W	P	A
29. Common Pochard	*Aythya ferina* (Linnaeus, 1758)	UC	M, W	P	A
30. Tufted Pochard	*Aythya fuligula* (Linnaeus, 1758)	UC	M, W	P	A
31. Ferruginous Pochard	*Aythya nyroca* (Guldenstadt, 1770)	R	M, W	P	A
32. Baer's Pochard	*Aythya baeri* (Radde, 1863)	R	M, W	P	A
33. Cotton Teal	*Nettapus coromandelianus* (Gmelin, 1789)	UC	M, W	P	A
34. Common Teal	*Anas crecca* Linnaeus, 1758	UC	M, W	P	A
35. Garganey	*Anas querquedula* Linnaeus, 1758	R	M, W	P	A
36. Gadwall	*Anas strepera* Linnaeus, 1758	UC	M, W	P	A
37. Eurasian Wigeon	*Anas penelope* Linnaeus, 1758	R	M, W	P	A
38. Northern Pintail	*Anas acuta* Linnaeus, 1758	UC	M, W	P	A
39. Northern Shoveller	*Anas clypeata* Linnaeus, 1758	UC	M, W	P	A
40. Brahminy Shelduck	*Tadorna ferruginea* (Pallas, 1764)	R	M, W	P	A
Family: Accipitridae					

Family and Common Name		Scientific Name	Abundance	Status	Food habit	Habitat
41.	Jerdon's Baza	*Aviceda jerdoni* (Blyth, 1842)	R	R	C	Ar, W
42.	Black Baza	*Aviceda leuphotes* (Dumont, 1820)	R	R	C	Ar, W
43.	Long-billed Vulture	*Gyps indicus* (Scopoli, 1786)	R	R	C	Ar, W
44.	Indian White-backed Vulture	*Gyps bengalensis* (Gmelin, 1788)	R	R	C	Ar, W
45.	Cinereous Vulture	*Aegypius monachus* (Linnaeus, 1766)	R	R	C	Ar, W
46.	Red-headed Vulture	*Sarcogyps calvus* (Scopoli, 1786)	R	R	C	Ar, W
47.	Egyptian Vulture	*Neophron percnopterus* (Linnaeus, 1758)	R	R	C	Ar, W
48.	Short-toed Snake-Eagle	*Circaetus gallicus* (Gmelin, 1788)	R	R	C	Ar, W
49.	Pallid Harrier	*Circus macrourus* (S.G. Gmelin, 1770)	R	R	C	Ar, W
50.	Pied Harrier	*Circus melanoleucos* (Pennant, 1769)	R	R	C	Ar, W
51.	Western Marsh-Harrier	*Circus aeruginosus* (Linnaeus, 1758)	R	R	C	Ar, W
52.	Montagu's Harrier	*Circus pygargus* (Linnaeus, 1758)	R	R	C	Ar, W
53.	Oriental Honey-Buzzard	*Pernis ptilorhynchus* (Temminck, 1821)	UC	R	C	Ar
54.	Black-shouldered Kite	*Elanus caeruleus* (Desfontaines, 1789)	C	R	C	Ar, O
55.	Black Kite	*Milvus migrans* (Boddaert, 1783)	C	R	C	Ar, O
56.	Brahminy Kite	*Haliastur indus* (Boddaert, 1783)	UC	R	C	O
57.	Crested Serpent-Eagle	*Spilornis cheela* (Latham, 1790)	C	R	C	W
58.	Black Eagle	*Ictinaetus malayensis* (Temminck, 1822)	R	R	C	W
59.	Tawny Eagle	*Aquila rapax* (Temminck, 1828)	R	R	C	Ar, W
60.	Shikra	*Accipiter badius* (Gmelin, 1788)	C	R	C	W, O
61.	Besra Sparrowhawk	*Accipiter virgatus* (Temminck, 1822)	UC	R	C	W, O
62.	Eurasian Sparrowhawk	*Accipiter nisus* (Linnaeus, 1758)	R	R	C	Ar, W
63.	Crested Goshawk	*Accipiter trivirgatus* (Temminck, 1824)	R	R	C	Ar, W
64.	Changeable Hawk-Eagle	*Spizaetus cirrhatus* (Gmelin, 1788)	R	R	C	Ar, W
65.	Oriental Honey-Buzzard	*Pernis ptilorhynchus* (Temminck, 1821)	R	R	C	Ar, W

Family and Common Name		Scientific Name	Abundance	Status	Food habit	Habitat
66.	White-eyed Buzzard	*Butastur teesa* (Franklin, 1832)	R	R	C	W, O
67.	Common Buzzard	*Buteo buteo* Linnaeus, 1758	R	R	C	Ar, W
68.	Greater Spotted Eagle	*Aquila clanga* Pallas, 1811	R	R	C	Ar, W
69.	Steppe Eagle	*Aquila nipalensis* Hodgson, 1833	R	R	C	Ar, W
70.	Bonelli's Eagle	*Hieraaetus fasciatus* (Vieillot, 1822)	R	R	C	Ar, W
71.	Booted Eagle	*Hieraaetus pennatus* (Gmelin, 1788)	R	R	C	Ar, W
72.	Rufous-bellied Eagle	*Hieraaetus kienerii* (E. Geoffroy, 1835)	R	R	C	Ar, W
73.	Pallas's Fish-Eagle	*Haliaeetus leucoryphus* (Pallas, 1771)	R	R	C	Ar, W
74.	Greater Grey-headed Fish-Eagle	*Ichthyophaga ichthyaetus* (Hors eld, 1821)	R	R	C	Ar, W
Family: Pandionidae						
75.	Osprey	*Pandion haliaetus* (Linnaeus, 1758)	R	UC	P	Ar, A
Family: Falconidae						
76.	Collared Falconet	*Microhierax caerulescens* (Linnaeus, 1758)	R	R	C	W
77.	Common Kestrel	*Falco tinnunculus* Linnaeus, 1758	R	R	C	W
78.	Lesser Kestrel	*Falco naumanni* Fleischer, 1818	R	R	C	W
79.	Peregrine Falcon	*Falco peregrinus* Tunstall, 1771	R	R	C	W
80.	Laggar	*Falco jugger* J.E. Gray, 1834	R	R	C	W
81.	Eurasian Hobby	*Falco subbuteo* Linnaeus, 1758	R	R	C	W
Family: Phasianidae						
82.	Grey Francolin	*Francolinus pondicerianus* (Gmelin, 1789)	UC	R	I	W
83.	Black Francolin	*Francolinus francolinus* (Linnaeus, 1766)	R	R	I	W
84.	Jungle Bush-Quail	*Perdicula asiatica* (Latham, 1790)	UC	R	I	W
85.	Red Junglefowl	*Gallus gallus* (Linnaeus, 1758)	C	R	I	W
86.	Indian Peafowl	*Pavo cristatus* Linnaeus, 1758	C	R	I	W

Family and Common Name		Scientific Name	Abundance	Status	Food habit	Habitat
87.	Red Spurfowl	*Galloperdix spadicea* (Gmelin, 1789)	R	R	I	W
88.	Painted Spurfowl	*Galloperdix lunulata* (Valenciennes, 1825)	R	R	I	W
89.	Rain Quail	*Coturnix coromandelica* (Gmelin, 1789)	UC	R	I	O, S
90.	Jungle Bush-Quail	*Perdicula asiatica* (Latham, 1790)	UC	R	I	O,S
91.	Blue-breasted Quail	*Coturnix chinensis* (Linnaeus, 1766)	R	R	I	O, S
92.	Painted Bush-Quail	*Perdicula erythrorhyncha* (Sykes, 1832)	R	R	I	O, S
Family: Turnicidae						
93.	Small Buttonquail	*Turnix sylvatica* (Desfontaines, 1789)	C	R	I	W
94.	Common Buttonquail	*Turnix suscitator* (Gmelin, 1789)	UC	R	I	O, S
95.	Yellow-legged Buttonquail	*Turnix tanki* Blyth, 1843	UC	R	I	O, S
Family: Rallidae						
96.	White-breasted Waterhen	*Amaurornis phoenicurus* (Pennant, 1769)	C	R	P	A
97.	Common Moorhen	*Gallinula chloropus* (Linnaeus, 1758)	C	R	P	A
98.	Purple Moorhen	*Porphyrio porphyrio* (Linnaeus, 1758)	C	R	P	A
99.	Common Coot	*Fulica atra* Linnaeus, 1758	C	R	P	A
100.	Brown Crake	*Amaurornis akool* (Sykes, 1832)	C	R	P	A
101.	Ruddy-breasted Crake	*Porzana fusca* (Linnaeus, 1766)	R	R	P	A
102.	Watercock	*Gallicrex cinerea* (Gmelin, 1789)	R	R	P	A
103.	Blue-breasted Rail	*Gallirallus striatus Linnaeus, 1766*	R	R	P	A
Family: Jacanidae						
104.	Bronze-winged Jacana	*Metopidius indicus* (Latham, 1790)	C	R	P	A
105.	Pheasant-tailed Jacana	*Hydrophasianus chirurgus* (Scopoli, 1786)	C	R	P	A
Family:Glareolidae						
106.	Small Pratincole	*Glareola lactea* Temminck, 1820	R	R	I	A

Family and Common Name		Scientific Name	Abundance	Status	Food habit	Habitat
Family: Charadriidae						
107.	Little Ringed Plover	*Charadrius dubius* Scopoli, 1786	UC	M, W	P, I	A
108.	Kentish Plover	*Charadrius alexandrinus* Linnaeus, 1758	R	M, W	P, I	A
109.	Grey-headed Lapwing	*Vanellus cinereus* (Linnaeus, 1758)	R	M, W	P, I	A
110.	Yellow-wattled Lapwing	*Vanellus malabaricus* (Boddaert, 1783)	C	R	I	O
111.	Red-wattled Lapwing	*Vanellus indicus* (Boddaert, 1783)	C	R	I	O
Family:Rostratulidae						
112.	Greater Painted-Snipe	*Rostratula benghalensis* (Linnaeus, 1758)	R	M, W	P	A
Family: Scolopacidae						
113.	Common Snipe	*Gallinago gallinago* (Linnaeus, 1758)	UC	M, W	P	A
114.	Common Redshank	*Tringa tetanus* (Linnaeus, 1758)	UC	M, W	P	A
115.	Common Greenshank	*Tringa nebularia* (Gunner, 1767)	UC	M, W	P	A
116.	Green Sandpiper	*Tringa ochropus* Linnaeus, 1758	UC	M, W	P	A
117.	Common Sandpiper	*Actitis hypoleucos* Linnaeus, 1758	C	M, W	P	A
118.	Wood Sandpiper	*Tringa glareola* Linnaeus, 1758	UC	M, W	P	A
119.	Little Stint	*Calidris minuta* (Leisler, 1812)	UC	M, W	P	A
120.	Temminck's Stint	*Calidris temminckii* (Leisler, 1812)	UC	M, W	P	A
Family:Burhinidae						
121.	Stone-Curlew	*Burhinus oedicnemus* (Linnaeus, 1758)	R	R	C	O
Family: Recurvirostridae						
122.	Black-winged Stilt	*Himantopus himantopus* (Linnaeus, 1758)	UC	R	P	A
Family: Laridae						
123.	River Tern	*Sterna aurantia* J.E. Gray, 1831	UC	R	P	A
124.	Brown-headed Gull	*Larus brunnicephalus* Jerdon, 1840	R	M, W	P	A
125.	Gull-billed Tern	*Gelochelidon nilotica (Gmelin, 1789)*	R	M, W	P	A

Family and Common Name		Scientific Name	Abundance	Status	Food habit	Habitat
126.	Black-bellied Tern	*Sterna acuticauda* J.E. Gray, 1831	R	R	P	A
127.	Whiskered Tern	*Chlidonias hybridus (Pallas, 1811)*	R	M, W	P	A
Family: Columbidae						
128.	Blue Rock Pigeon	*Columba livia* Gmelin, 1789	C	R	G	W, O
129.	Purple Wood-Pigeon	*Columba punicea* Blyth, 1842	UC	R	G	W, O
130.	Oriental Turtle-Dove	*Streptopelia orientalis* (Latham, 1790)	UC	R	G	W, O
131.	Spotted Dove	*Streptopelia chinensis* (Scopoli, 1786)	C	R	G	W, O
132.	Eurasian Collared-Dove	*Streptopelia decaocto* (Frivaldszky, 1838)	C	R	G	W, O
133.	Little Brown Dove	*Streptopelia senegalensis* (Linnaeus, 1766)	C	R	G	W, O
134.	Red Collared-Dove	*Streptopelia tranquebarica* (Hermann, 1804)	UC	R	G	W, O
135.	Emerald Dove	*Chalcophaps indica* (Linnaeus, 1758)	UC	R	G	W, O
136.	Thick-billed Green-Pigeon	*Treron curvirostra* (Gmelin, 1789)	UC	R	G	W, O
137.	Yellow-legged Green-Pigeon	*Treron phoenicoptera* (Latham, 1790)	UC	R	G	W, O
138.	Orange-breasted Green-Pigeon	*Treron bicincta* (Jerdon, 1840)	R	R	G	W, O
139.	Pompadour Green-Pigeon	*Treron pompadora* (Gmelin, 1789)	R	R	G	W, O
140.	Pale-capped Pigeon	*Columba punicea*(Gmelin, 1789)	UC	R	G	W, O
141.	Green Imperial-Pigeon	*Ducula aenea* (Linnaeus, 1766)	UC	R	G	W, O
142.	Mountain Imperial-Pigeon	*Ducula badia* (Raf es, 1822)	UC	R	G	W, O
Family: Pteroclididae						
143.	Chestnut-bellied Sandgrouse	*Pterocles exustus* Temminck, 1825	R	R	G	O
Family: Psittacidae						
144.	Rose-ringed Parakeet	*Psittacula krameri* (Scopoli, 1769)	C	R	F	W
145.	Alexandrine Parakeet	*Psittacula eupatria* (Linnaeus, 1766)	C	R	F	W
146.	Plum-headed Parakeet	*Psittacula cyanocephala* (Linnaeus, 1766)	C	R	F	W

Family and Common Name	Scientific Name	Abundance	Status	Food habit	Habitat
147. Indian Hanging-Parrot	*Loriculus vernalis* (Sparrman, 1787)	UC	R	F	W
Family: Cuculidae					
148. Brainfever Bird	*Hierococcyx varius* (Vahl, 1797)	UC	R	I	W
149. Large Hawk-Cuckoo	*Hierococcyx sparverioides* (Vigors, 1832)	UC	R	I	W
150. Pied Crested Cuckoo	*Clamator jacobinus* (Boddaert, 1783)	UC	R	I	W
151. Red-winged Crested Cuckoo	*Clamator coromandus (Linnaeus, 1766)*	UC	R	I	W
152. Indian Cuckoo	*Cuculus micropterus* Gould, 1838	UC	R	I	W
153. Oriental Cuckoo	*Cuculus saturatus* Blyth, 1843	UC	R	I	W
154. Lesser Cuckoo	*Cuculus poliocephalus* Latham, 1790	UC	R	I	W
155. Indian Plaintive Cuckoo	*Cacomantis passerinus* (Vahl, 1797)	R	M, S	I	W
156. Banded Bay Cuckoo	*Cacomantis sonneratii* (Latham, 1790)	R	R	I	W
157. Drongo Cuckoo	*Surniculus lugubris (Horsfield, 1821)*	R	R	I	W
158. Asian Koel	*Eudynamys scolopacea* (Linnaeus, 1758)	C	R	I, F	W
159. Sirkeer Malkoha	*Phaenicophaeus leschenaultii* (Lesson, 1830)	UC	R	I, C	W
160. Small Green-billed Malkoha	*Phaenicophaeus viridirostris* (Jerdon, 1840)	UC	R	I, C	W
161. Large Green-billed Malkoha	*Phaenicophaeus tristis* (Lesson, 1830)	UC	R	I, C	W
162. Lesser Coucal	*Centropus bengalensis* (Gmelin, 1788)	R	R	I, C	W
163. Greater Coucal	*Centropus sinensis* (Stephens, 1815)	C	R	I, C	O
Family:Tytonidae					
164. Barn Owl	*Tyto alba* (Scopoli, 1769)	UC	R	C	W
Family: Strigidae					
165. Brown Fish-Owl	*Ketupa zeylonensis* (Gmelin, 1788)	UC	R	C	W
166. Collared Scops-Owl	*Otus bakkamoena* Pennant, 1769	R	R	C	W
167. Oriental Scops-Owl	*Otus sunia* (Hodgson, 1836)	R	R	C	W
168. Brown Hawk-Owl	*Ninox scutulata* (Raf es, 1822)	R	R	C	W

Family and Common Name	Scientific Name	Abundance	Status	Food habit	Habitat
169. Short-eared Owl	*Asio flammeus* (Pontoppidan, 1763)	R	R	C	W
170. Eurasian Eagle-Owl	*Bubo bubo* (Linnaeus, 1758)	UC	R	C	W
171. Forest Eagle-Owl	*Bubo nipalensis* Hodgson, 1836	UC	R	C	W
172. Brown Wood-Owl	*Strix leptogrammica* Temminck, 1831	R	R	C	W
173. Mottled Wood-Owl	*Strix ocellata* (Lesson, 1839)	R	R	C	W
174. Jungle Owlet	*Glaucidium radiatum* (Tickell, 1833)	C	R	C	W
175. Spotted Owlet	*Athene brama* (Temminck, 1821)	C	R	C	W
Family: Caprimulgidae					
176. Indian Jungle Nightjar	*Caprimulgus indicus* Latham, 1790	C	R	I	W
177. Large-tailed Nightjar	*Caprimulgus macrurus* Hors eld, 1821	UC	R	I	W
178. Common Indian Nightjar	*Caprimulgus asiaticus* Latham, 1790	C	R	I	W
179. Franklin's Nightjar	*Caprimulgus affinis* Hors eld, 1821	UC	R	I	W
Family: Apodidae					
180. Asian Palm-Swift	*Cypsiurus balasiensis* (J.E. Gray, 1829)	C	R	I	O, S
181. House Swift	*Apus affinis* (J.E. Gray, 1830)	C	R	I	O, S
182. White-rumped Needletail-Swift	*Zoonavena sylvatica (Tickell, 1846)*	UC	R	I	O, S
Family:Hemiprocnidae					
183. Crested Tree-Swift	*Hemiprocne coronata* (Tickell, 1833)	UC	R	I	O, S
Family: Alcedinidae					
184. Small Blue King sher	*Alcedo atthis* (Linnaeus, 1758)	C	R	P	A
185. White-breasted King sher	*Halcyon smyrnensis* (Linnaeus, 1758)	C	R	P	A
186. Lesser Pied King sher	*Ceryle rudis* (Linnaeus, 1758)	UC	R	P	A
187. Stork-billed King sher	*Halcyon capensis* (Linnaeus, 1766)	R	M, W	P	A
Family: Meropidae					

Family and Common Name		Scientific Name	Abundance	Status	Food habit	Habitat
188.	Small Bee-eater	*Merops orientalis* Latham, 1801	C	R	I	O
189.	Blue-tailed Bee-eater	*Merops philippinus* Linnaeus, 1766	UC	R	I	O
190.	Chestnut-headed Bee-eater	*Merops leschenaulti* Vieillot, 1817	UC	R	I	W, O
191.	Blue-bearded Bee-eater	*Nyctyornis athertoni* (Jardine & Selby, 1828)	UC	R	I	W
Family: Coraciidae						
192.	Indian Roller	*Coracias benghalensis* (Linnaeus, 1758)	C	R	I	O
Family: Upupidae						
193.	Common Hoopoe	*Upupa epops* Linnaeus, 1758	UC	R	I	O
Family: Bucerotidae						
194.	Indian Grey Hornbill	*Ocyceros birostris* (Scopoli, 1786)	R	R	F	W
195.	Malabar Pied Hornbill	*Anthracoceros coronatus* (Boddaert, 1783)	R	R	F	W
196.	Oriental Pied Hornbill	*Anthracoceros albirostris* (Shaw, 1808)	R	R	F	W
Family: Capitonidae						
197.	Coppersmith Barbet	*Megalaima haemacephala* (P.L.S. Müller, 1776)	C	R	I, F	W
198.	Brown-headed Barbet	*Megalaima zeylanica* (Gmelin, 1788)	C	R	I, F	W
199.	Lineated Barbet	*Megalaima lineata* (Vieillot, 1816)	UC	R	I, F	W
200.	Blue-throated Barbet	*Megalaima asiatica* (Latham, 1790)	UC	R	I, F	W
Family: Trogonidae						
201.	Malabar Trogon	*Harpactes fasciatus* (Pennant, 1769)	UC	R	I, F	W
Family: Picidae						
202.	Eurasian Wryneck	*Jynx torquilla* Linnaeus, 1758	UC	M, W	I	O
203.	Speckled Piculet	*Picumnus innominatus* Burton, 1836	UC	R	I	W
204.	Fulvous-breasted Pied Woodpecker	*Dendrocopos macei* (Vieillot, 1818)	UC	R	I	W

Family and Common Name		Scientific Name	Abundance	Status	Food habit	Habitat
205.	Brown-capped Pygmy Woodpecker	*Dendrocopos nanus* (Vigors, 1832)	UC	R	I	W
206.	Yellow-fronted Pied Woodpecker	*Dendrocopos mahrattensis* (Latham, 1801)	UC	R	I	W
207.	Lesser Golden-backed Woodpecker	*Dinopium benghalense* (Linnaeus, 1758)	UC	R	I	W
208.	Greater Golden-backed Woodpecker	*Chrysocolaptes lucidus* (Scopoli, 1786)	UC	R	I	W
209.	Black-shouldered Woodpecker	*Chrysocolaptes festivus* (Boddaert, 1783)	UC	R	I	W
210.	Rufous Woodpecker	*Celeus brachyurus* (Vieillot, 1818)	UC	R	I	W
211.	Small Yellow-naped Woodpecker	*Picus chlorolophus* Vieillot, 1818	UC	R	I	W
212.	Large Yellow-naped Woodpecker	*Picus flavinucha* Gould, 1834	UC	R	I	W
213.	Heart-spotted Woodpecker	*Hemicircus canente* (Lesson, 1830)	UC	R	I	W
214.	Great Black Woodpecker	*Dryocopus javensis* (Hors eld, 1821)	UC	R	I	W
Family: Alaudidae						
215.	Red-winged Bush-Lark	*Mirafra erythroptera* Blyth, 1845	UC	R	I	O
216.	Bengal Bush-Lark	*Mirafra assamica* Hors eld, 1840	UC	R	I	O
217.	Jerdon's Bush-Lark	*Mirafra affinis* Blyth, 1845	UC	R	I	O
218.	Rufous-tailed Finch-Lark	*Ammomanes phoenicurus* (Franklin, 1831)	UC	R	I	O
219.	Eastern Skylark	*Alauda gulgula* Franklin, 1831	UC	R	I	O
220.	Ashy-crowned Sparrow-Lark	*Eremopterix grisea* (Scopoli, 1786)	C	R	I	O
Family: Hirundinidae						
221.	Common Swallow	*Hirundo rustica* Linnaeus, 1758	UC	R	I	O
222.	Red-rumped Swallow	*Hirundo daurica* Linnaeus, 1771	UC	R	I	O
223.	Dusky Crag-Martin	*Hirundo concolor* Sykes, 1833	R	R	I	O
224.	Northern House-Martin	*Delichon urbica* (Linnaeus, 1758)	R	R	I	O

Family and Common Name		Scientific Name	Abundance	Status	Food habit	Habitat
225.	Wire-tailed Swallow	*Hirundo smithii* Leach, 1818	UC	R	I	O
226.	Streak-throated Swallow	*Hirundo fluvicola* Blyth, 1855	UC	R	I	O
Family: Motacillidae						
227.	White Wagtail	*Motacilla alba* Linnaeus, 1758	UC	M, W	I	O, A
228.	Large Pied Wagtail	*Motacilla maderaspatensis* Gmelin, 1789	UC	M, W	I	O, A
229.	Citrine Wagtail	*Motacilla citreola* Pallas, 1776	UC	M, W	I	O, A
230.	Yellow Wagtail	*Motacilla flava* Linnaeus, 1758	UC	M, W	I	O, A
231.	Grey Wagtail	*Motacilla cinerea* Tunstall, 1771	UC	M, W	I	O, A
232.	Forest Wagtail	*Dendronanthus indicus* (Gmelin, 1789)	UC	M, W	I	O, A
233.	Paddy eld Pipit	*Anthus rufulus* Vieillot, 1818	C	R	I	O
234.	Oriental Tree Pipit	*Anthus hodgsoni* Richmond, 1907	UC	R	I	O
235.	Eurasian Tree Pipit	*Anthus trivialis* (Linnaeus, 1758)	UC	R	I	O
236.	Blyth's Pipit	*Anthus godlewskii* (Taczanowski, 1876)	UC	R	I	O
Family: Campephagidae						
237.	Large Cuckoo-Shrike	*Coracina macei* (Lesson, 1830)	UC	R	I	W
238.	Black-headed Cuckoo-Shrike	*Coracina melanoptera* (Rüppell, 1839)	UC	R	I	W
239.	Black-winged Cuckoo-Shrike	*Coracina melaschistos* (Hodgson, 1836)	UC	R	I	W
240.	Rosy Minivet	*Pericrocotus roseus* (Vieillot, 1818)	UC	R	I	W
241.	Small Minivet	*Pericrocotus cinnamomeus* (Linnaeus, 1766)	UC	R	I	W
242.	Scarlet Minivet	*Pericrocotus flammeus* (Forster, 1781)	C	R	I	W
243.	Common Woodshrike	*Tephrodornis pondicerianus* (Gmelin, 1789)	UC	R	I	W
244.	Large Woodshrike	*Tephrodornis gularis* (Raf es, 1822)	UC	R	I	W
245.	Pied Flycatcher-Shrike	*Hemipus picatus* (Sykes, 1832)	R	R	I	W
Family: Pycnonotidae						
246.	White-browed Bulbul	*Pycnonotus luteolus* (Lesson, 1841)	UC	R	I	W

Family and Common Name		Scientific Name	Abundance	Status	Food habit	Habitat
247.	Black-crested Bulbul	*Pycnonotus melanicterus* (Gmelin, 1789)	C	R	I	W
248.	Red-whiskered Bulbul	*Pycnonotus jocosus* (Linnaeus, 1758)	C	R	I	W
249.	Red-vented Bulbul	*Pycnonotus cafer* (Linnaeus, 1766)	C	R	I	W
Family: Irenidae						
250.	Common Iora	*Aegithina tiphia* (Linnaeus, 1758)	C	R	I	W
251.	Jerdon's Chloropsis	*Chloropsis cochinchinensis* (Gmelin, 1788)	C	R	I	W
252.	Gold-fronted Chloropsis	*Chloropsis aurifrons* (Temminck, 1829)	C	R	I	W
253.	Asian Fairy-Bluebird	*Irena puella* (Latham, 1790)	UC	R	I	W
Family: Pittidae						
254.	Indian Pitta	*Pitta brachyura* (Linnaeus, 1766)	UC	M, S	I	W
Family: Laniidae						
255.	Brown Shrike	*Lanius cristatus* Linnaeus, 1758	C	M, W	I	O
256.	Bay-backed Shrike	*Lanius vittatus* Valenciennes, 1826	UC	R	I	O
257.	Rufous-backed Shrike	*Lanius schach* Linnaeus, 1758	UC	R	I	W
258.	Grey-backed Shrike	*Lanius tephronotus* (Vigors, 1831)	UC	R	I	W, O
Family: Turdinae						
259.	Blue Rock-Thrush	*Monticola solitarius* (Linnaeus, 1758)	UC	R	I	R, W
260.	Blue-headed Rock-Thrush	*Monticola cinclorhynchus* (Vigors, 1832)	UC	R	I	R, W
261.	Malabar Whistling-Thrush	*Myiophonus horsfieldii* (Vigors, 1831)	UC	R	I	R, W
262.	Tickell's Thrush	*Turdus unicolor* Tickell, 1833	UC	R	I	R, W
263.	Orange-headed Thrush	*Zoothera citrina* (Latham, 1790)	UC	R	I	R, W
264.	Scaly Thrush	*Zoothera dauma* (Latham, 1790)	UC	R	I	R, W
265.	Eurasian Blackbird	*Turdus merula* Linnaeus, 1758	R	R	I	S, W, O
266.	Indian Robin	*Saxicoloides fulicata* (Linnaeus, 1776)	C	R	I	O

Family and Common Name		Scientific Name	Abundance	Status	Food habit	Habitat
267.	Oriental Magpie-Robin	*Copsychus saularis* (Linnaeus, 1758)	C	R	I	O
268.	White-rumped Shama	*Copsychus malabaricus* (Scopoli, 1786)	C	R	I	W
269.	Black Redstart	*Phoenicurus ochruros* (Gmelin, 1774)	R	M, W	I	R, W
270.	Bluethroat	*Luscinia svecica* (Linnaeus, 1758)	R	R	I	O, W
271.	Common Stonechat	*Saxicola torquata* (Linnaeus, 1766)	R	R	I	O, W
272.	White-tailed Stonechat	*Saxicola leucura* (Blyth, 1847)	R	R	I	O, W
273.	Pied Bushchat	*Saxicola caprata* (Linnaeus, 1766)	UC	R	I	O, W
274.	Desert Wheatear	*Oenanthe deserti* (Temminck, 1825)	R	M	I	O
275.	Variable Wheatear	*Oenanthe picata* (Blyth, 1847)	R	M	I	O
Family: Timaliinae						
276.	Spotted Babbler	*Pellorneum ruficeps* Swainson, 1832	UC	R	I	W
277.	Indian Scimitar-Babbler	*Pomatorhinus horsfieldii* Sykes, 1832	R	R	I	W
278.	Yellow-eyed Babbler	*Chrysomma sinense* (Gmelin, 1789)	C	R	I	W, O
279.	Jungle Babbler	*Turdoides striatus* (Dumont, 1823)	C	R	I	W, O
280.	Common Babbler	*Turdoides caudatus* (Dumont, 1823)	C	R	I	W, O
281.	Quaker Tit-Babbler	*Alcippe poioicephala* (Jerdon, 1844)	R	R	I	W
282.	Rufous-capped Babbler	*Stachyris ruficeps* Blyth, 1847	R	R	I	W, O
283.	Yellow-breasted Babbler	*Macronous gularis* (Hors eld, 1822)	R	R	I	W, O
284.	Rufous-bellied Babbler	*Dumetia hyperythra* (Franklin, 1831)	R	R	I	W, O
Family: Sylviinae						
285.	Jungle Prinia	*Prinia sylvatica* Jerdon, 1840	UC	R	I	O, S
286.	Franklin's Prinia	*Prinia hodgsonii* Blyth, 1844	UC	R	I	O, S
287.	Ashy Prinia	*Prinia socialis* Sykes, 1832	UC	R	I	O, S
288.	Plain Prinia	*Prinia inornata* Sykes, 1832	UC	R	I	O, S

Family and Common Name		Scientific Name	Abundance	Status	Food habit	Habitat
289.	Common Tailorbird	*Orthotomus sutorius* (Pennant, 1769)	C	R	I	W, O, S
290.	Common Chiffchaff	*Phylloscopus collybita* (Vieillot, 1817)	UC	M, W	I	O, S
291.	Greenish Leaf-Warbler	*Phylloscopus trochiloides* (Sundevall, 1837)	C	R	I	W, O, S
292.	Hume's Warbler	*Phylloscopus humei* (Brooks, 1878)	R	R	I	W, O, S
293.	Blanford's Bush-Warbler	*Cettia pallidipes* (Blanford, 1872)	UC	R	I	O, S
294.	Spotted Bush-Warbler	*Bradypterus thoracicus* (Blyth, 1845)	UC	R	I	O, S
295.	Thick-billed Warbler	*Acrocephalus aedon* (Pallas, 1776)	UC	R	I	O, S
296.	Blyth's Reed-Warbler	*Acrocephalus dumetorum* Blyth, 1849	UC	R	I	O, S
297.	Paddy eld Warbler	*Acrocephalus agricola* (Jerdon, 1845)	UC	R	I	O, S
298.	Indian Great Reed-Warbler	*Acrocephalus stentoreus* (Hemprich & Ehrenberg, 1833)	UC	R	I	O, S
299.	Bristled Grass-Warbler	*Chaetornis striatus* (Jerdon, 1841)	R	R	I	O, S
300.	Streaked Fantail-Warbler	*Cisticola juncidis* (Ra nesque, 1810)	C	R	I	O, S
301.	Orphean Warbler	*Sylvia hortensis* (Gmelin, 1789)	R	R	I	O, S
302.	Tickell's Warbler	*Phylloscopus affinis* (Tickell, 1833)	R	R	I	O, S
Family: Muscicapinae						
303.	Red-throated Flycatcher	*Ficedula parva* (Bechstein, 1792)	R	M, W	I	W, O
304.	Tickell's Blue-Flycatcher	*Cyornis tickelliae* Blyth, 1843	UC	M, W	I	W, O
305.	Asian Brown Flycatcher	*Muscicapa dauurica* Pallas, 1811	UC	M, W	I	W, O
306.	Brown-breasted Flycatcher	*Muscicapa muttui* (Layard, 1854)	UC	M, W	I	W, O
307.	Ultramarine Flycatcher	*Ficedula superciliaris* (Jerdon, 1840)	UC	M, W	I	W, O
308.	Little Pied Flycatcher	*Ficedula westermanni* (Sharpe, 1888)	UC	R	I	W, O
309.	Verditer Flycatcher	*Eumyias thalassina* (Swainson, 1838)	UC	M, W	I	W, O
310.	Brooks's Flycatcher	*Cyornis poliogenys* Brooks, 1879	UC	R	I	W, O
311.	Blue-throated Flycatcher	*Cyornis rubeculoides* (Vigors, 1831)	UC	R	I	W, O

Family and Common Name		Scientific Name	Abundance	Status	Food habit	Habitat
312.	Grey-headed Flycatcher	*Culicicapa ceylonensis* (Swainson, 1820)	UC	M, W	I	W, O
Family: Monarchinae						
313.	Asian Paradise-Flycatcher	*Terpsiphone paradisi* (Linnaeus, 1758)	UC	M, S	I	W
314.	Black-naped Monarch	*Hypothymis azurea* (Boddaert, 1783)	C	R	I	W, S
Family: Rhipidurinae						
315.	White-browed Fantail	*Rhipidura aureola* Lesson, 1830	C	R	I	W
316.	White-throated Fantail-Flycatcher	*Rhipidura albicollis* (Vieillot, 1818)	C	R	I	W
Family: Dicaeidae						
317.	Thick-billed Flowerpecker	*Dicaeum agile* (Tickell, 1833)	C	R	I, F	W
318.	Tickell's Flowerpecker	*Dicaeum erythrorhynchos* (Latham, 1790)	UC	R	I, F	W
Family: Nectariniidae						
319.	Purple Sunbird	*Nectarinia asiatica* (Latham, 1790)	C	R	N	W
320.	Purple-rumped Sunbird	*Nectarinia zeylonica* (Linnaeus, 1766)	C	R	N	W
321.	Crimson Sunbird	*Aethopyga siparaja* (Raf es, 1822)	UC	R	N	W
322.	Little Spiderhunter	*Arachnothera longirostra* (Latham, 1790)	R	R	N	W
Family: Zosteropidae						
323.	Oriental White-eye	*Zosterops palpebrosus* (Temminck1824)	C	R	N	W
Family: Fringillidae						
324.	Common Rose nch	*Carpodacus erythrinus* (Pallas, 1770)	R	M, W	G	W, O
Family:Emberizinae						
325.	Crested Bunting	*Melophus lathami* (Gray, 1831)	R	R	G	W, O
326.	Grey-necked Bunting	*Emberiza buchanani* Blyth, 1845	R	M	G	O, S
Family: Estrildidae						

Family and Common Name		Scientific Name	Abundance	Status	Food habit	Habitat
327.	Red Munia	*Amandava amandava* (Linnaeus, 1758)	UC	R	G	O, S
328.	White-throated Munia	*Lonchura malabarica* (Linnaeus, 1758)	C	R	G	O, S
329.	White-rumped Munia	*Lonchura striata* (Linnaeus, 1766)	UC	R	G	O, S
330.	Spotted Munia	*Lonchura punctulata* (Linnaeus, 1758)	C	R	G	O, S
331.	Black-headed Munia	*Lonchura malacca* (Linnaeus, 1766)	UC	R	G	O, S
332.	Black-throated Munia	*Lonchura kelaarti* (Jerdon, 1863)	UC	R	G	O, S
Family: Passerinae						
333.	House Sparrow	*Passer domesticus* (Linnaeus, 1758)	C	R	G	O
334.	Yellow-throated Sparrow	*Petronia xanthocollis* (Burton, 1838)	UC	R	G	O
Family: Ploceinae						
335.	Baya Weaver	*Ploceus philippinus* (Linnaeus, 1766)	UC	R	G	O
336.	Black-breasted Weaver	*Ploceus benghalensis* (Linnaeus, 1758)	UC	R	G	O
337.	Streaked Weaver	*Ploceus manyar* (Hors eld, 1821)	UC	R	G	O
Family: Sturnidae						
338.	Brahminy Starling	*Sturnus pagodarum*(Gmelin, 1789)	UC	R	I, F	O, W, S
339.	Grey-headed Starling	*Sturnus malabaricus* (Gmelin, 1789)	UC	R	I, F	O, W, S
340.	Asian Pied Starling	*Sturnus contra* Linnaeus, 1758	C	R	I, F	O, W, S
341.	Rosy Starling	*Sturnus roseus* (Linnaeus, 1758)	UC	R	I, F	O, W, S
342.	Common Myna	*Acridotheres tristis* (Linnaeus, 1766)	C	R	I, F	O, W, S
343.	Bank Myna	*Acridotheres ginginianus* (Latham, 1790)	R	M	I, F	O, S
344.	Jungle Myna	*Acridotheres fuscus* (Wagler, 1827)	C	R	I, F	O, W, S
345.	Common Hill-Myna	*Gracula religiosa* Linnaeus, 1758	UC	R	I, F	O, W
Family: Oriolidae						
346.	Eurasian Golden Oriole	*Oriolus oriolus* (Linnaeus, 1758)	UC	R	I	O, W, S

Family and Common Name		Scientific Name	Abundance	Status	Food habit	Habitat
347.	Black-headed Oriole	*Oriolus xanthornus* (Linnaeus, 1758)	C	R	I	O, W, S
348.	Black-naped Oriole	*Oriolus chinensis* Linnaeus, 1766	UC	R	I	O, W, S
Family: Dicruridae						
349.	Black Drongo	*Dicrurus macrocercus* Vieillot, 1817	C	R	I,	O, W, S
350.	Ashy Drongo	*Dicrurus leucophaeus* Vieillot, 1817	UC	R	I	W, S
351.	White-bellied Drongo	*Dicrurus caerulescens* (Linnaeus, 1758)	UC	R	I	W, S
352.	Bronzed Drongo	*Dicrurus aeneus* Vieillot, 1817	UC	R	I	W, S
353.	Spangled Drongo	*Dicrurus hottentottus* (Linnaeus, 1766)	UC	R	I	W, S
354.	Greater Racket-tailed Drongo	*Dicrurus paradiseus* (Linnaeus, 1766)	UC	R	I	W
Family: Corvidae						
355.	Indian Treepie	*Dendrocitta vagabunda* (Latham, 1790)	C	R	I, C	W, S
356.	Grey Treepie	*Dendrocitta formosae* Swinhoe, 1863	UC	R	I, C	W, S
357.	House Crow	*Corvus splendens* Vieillot, 1817	C	R	C	O, W, S
358.	Jungle Crow	*Corvus macrorhynchos* Wagler, 1827	UC	R	C	O, W, S
Family: Artamidae						
359.	Ashy Woodswallow	*Artamus fuscus* Vieillot, 1817	UC	R	I	O, W, S
Family: Paridae						
360.	Great Tit	*Parus major* Linnaeus, 1758	UC	R	I	O, W, S
361.	Black-lored Yellow Tit	*Parus xanthogenys* Vigors, 1831	UC	R	I	O, W, S
Family: Sittidae						
362.	Velvet-fronted Nuthatch	*Sitta frontalis* Swainson, 1820	C	R	I	W
363.	Chestnut-bellied Nuthatch	*Sitta castanea* Lesson, 1830	C	R	I	W

Table 8.2: List of globally threatened bird species in Similipal and their status (IUCN/Birdlife International Red Data List 2014).

Common Name	Scientific Name	IUCN Status
Black Ibis	*Pseudibis papillosa*	Near Threatened
Ferruginous Pochard	*Aythya nyroca*	Near Threatened
Baer's Pochard	*Aythya baeri*	Vulnerable
River Tern	*Sterna aurantia*	Near Threatened
Black-bellied Tern	*Sterna acuticauda*	Endangered
Painted Stork	*Mycteria leucocephala* (Pennant, 1769)	Near Threatened
Oriental White Ibis	*Threskiornis melanocephalus* (Latham, 1790)	Near Threatened
Greater Spotted Eagle	*Aquila clanga*	Vulnerable
Cinereous Vulture	*Aegypius monachus*	Near Threatened
Long-billed Vulture	*Gyps indicus* (Scopoli, 1786)	Critically Endangered
Indian White-backed Vulture	*Gyps bengalensis* (Gmelin, 1788)	Critically Endangered
Red-headed Vulture	*Sarcogyps calvus*	Critically Endangered
Pallas's Fish-Eagle	*Haliaeetus leucoryphus* (Pallas, 1771)	Vulnerable
Malabar Pied Hornbill	*Anthracoceros coronatus*	Near Threatened
Alexandrine Parakeet	*Psittacula eupatria*	Near Threatened
Pale-capped Pigeon	*Columba punicea*	Vulnerable
Bristled Grass-Warbler	*Chaetornis striatus* (Jerdon, 1841)	Vulnerable

Table 8.3 : Waterfowl population in three different divisions of similipal during 2012.

SL. No.	Common name	Scientific name	Baripada Division	Rairangpur Division	Karanjia Division
Family Podicipedidae					
1	Little Grebe	*Tachybaptua ruflcollis*	197	45	42
2	Great crested Grebe	*Podiceps cristatus*	23	23	11
Family: Phalacrocoracidae					
3	Indian Shag	*Phalacrocorax fuscicollis*	19		34
4	Little cormorant	*Phalacrocorax niger*	480	27	12
Family: Ardeidae					
5	Indian Pond Heron	*Ardeola grayii*	159	19	11
6	Cattle Egret	*Bubulcus ibis*	258	19	14
7	Little Egret	*Egretta garzetta*	188	17	17
8	Intermediate Egret	*Egretta intermedia*	13	4	10
9	Great Egret	*Egretta alba*	2		
10	Purple Heron	*Ardea purpurea*	1	0	1
Family: Ciconiidae					
11	Asian Openbill	*Anastomus oscitans*	29		
12	White-necked Stork	*Ciconia episcopus*	1		
13	Lesser Whistling Duck	*Dendrocygna javanica*	480	70	4
14	Indian Cotton teal	*Nettapus coromandeltanus*	78		4
15	Eurasian Wigeon	*Anes Penelope*	45		
16	Gadwall	*Anes strepera*	231	600	70
17	Northern Pintail	*Anes acuta*	6		

SL. No.	Common name	Scientific name	Baripada Division	Rairangpur Division	Karanjia Division
18	Red-crested Pochard	*Rhodonessa ruffian*	135	340	60
19	Common Pochard	*Aythya farina*		330	70
20	Ferruginous Duck	*Aythya nyroca*	22		
21	Tufted duck	*Aythya fuligula*	432	80	0
22	White Breasted Water Hen	*Amaurornis phoenicurus*	3		
23	Moorhen	*Gallinula chloropus*			25
24	Purple Swamphen	*Porphyrio porphyrio*	2		9
25	Common Coot	*Fulica atra*	83	120	0
26	Phesant-tailed Jacana	*Hydrophasianus chirugus*	60	0	8
27	Bronze-winged Jacana	*Metopidius indicus*	83		2
28	Yellow-wattled Lapwing	*Vanellus malabaricus*	5		6
29	Grey Headed Lapwing	*Vanellus cinereus*	110		
30	Red-wattled Lapwing	*Vanellus indicus*	9		6
31	Little Ringed Plover	*Charadrius dubius*	3		
32	Greenshank	*Tringa nebularia*			3
33	Wood Sandpiper	*Tringa glareola*			1
34	Common Sand Piper	*Aclitis hypoleucos*	9	5	23
36	Common Snipe	Gallinago megala	8		4
37	Jack Snipe	Lymnocryptes minimus	31		
38	Little Stint	*Calidris minuta*		5	0

Chapter-9

STATUS OF MAMMALS IN SIMILIPAL

Mammals are extremely important for functioning and well balancing of ecosystems. They share responsibility of pollinating plants and dispersing seeds; they are both predator and prey and can have immense effects on the structure and composition of vegetation, plant productivity and nutrient cycling. Many mammal species are experiencing greater population declines than any of the other vertebrate groups.

Estimating the mammals' abundance of populations is of central importance in conservation. This basic information is critical to define conservation priorities (Mackenzie, 2005), or to evaluate effectiveness of management policies. In the tropical forest of India, medium and large terrestrial mammals are the focus of most biodiversity monitoring programs (Karanth et al., 2009) mainly due to their vulnerability to multiple human-driven threats and because of their importance in forest system dynamics (Karantha et al., 2010). However, monitoring such species in large remote forest areas is difficult due their elusive behavior and low abundances (Datta et al., 2008). Several methods have been used to estimate the absolute abundance of mammals, such as capture-mark-recapture (Shanker, 2000), line transect (Jathana et al., 2003), aerial counts (Mourão et al., 2000), defecation rates (Plumptre and Harris, 1995) and footprints measurements (Sagar and Singh 1990,

Sharma et al., 2005). Over the past decades camera traps have become an important tool for monitoring rare, cryptic species in a wide range environment (Karanth and Nichols, 1998; Cutler and Swann, 1999). Camera traps opened up an invaluable opportunity to monitor wildlife populations non-invasively, even in logistically challenging environmental condition (Kays and Slauson, 2008; O'Connell et al., 2011). Their ability to quickly accumulate data over large areas with relatively little effort makes them an ideal tool for wildlife monitoring and biodiversity assessment.

As estimating abundance and distribution of species is fundamental to ecological studies, it requires efficient methodology. Reliable estimates of population size are often necessary before any management activity undertaken. However, accurate estimates are often difficult to acquire. In line transect method sighting of animals can be used to estimate species densities and abundances (Buckland et al., 2001). However, the small number of sightings (Jathana et al., 2003) frequently leads to poor estimates of abundance and occurrence of mammals in tropical forest, especially for nocturnal and less abundant species, like small and large body carnivores (Tobler et al., 2008). Nonetheless, these species are often among the most vulnerable to anthropogenic disturbances (Madhusudan and Mishra, 2003). Camera traps acknowledge important tools for monitoring nocturnal and cryptic species, and further development of their use has led to population estimation of natural marked animals by means of well consolidated capture-recaptures models. (Karanth, 1995, Karanth and Nichols, 1998). However, the most reliable abundance estimation method capture-recapture has difficult to achieve at larger spatial scales (Mackenzie et al., 2002; Pollock et al., 2002), and it is only possible to identify individual natural marked animals. Therefore, for the majority of tropical animals, including ungulate, bears and other small mammals, it is not possible to identify individual with confidence. In this scenario, another approach has been proposed to apply abundance estimations with camera trapping data for a wider range species. In a simple and direct way, trapping rates (photographs/trapping effort) became widely used method in most studies (Carbone et al., 2001; Trolle and Kéry, 2005). Its usage as a relative index of abundance has been supported by evidence of the existence of significant correlation between trapping rates and independent density estimations in a number of species (Carbone et al., 2001; O'Brien et al., 2003). The use of relative abundance index (RAI) based on camera trap encounter rates for ecological studies is controversial particularly when comparing between species as a large number of variables (e.g. body size, average group size, behavior) are likely to affect trapping rates and detection probability and thus confound the relationship with actual abundance (Carbone et al., 2001; Jennelle et al., 2002; Treves et al., 2010). However, there is increasing evidence for a linear relationship between RAI and abundance estimated through more rigorous methodologies (Rovero and Marshall, 2009). Therefore, taking into account the caveats above, we estimated medium to large sized mammal abundances through RAI among fixed camera locations within our study area.

Methods of Study of Mammals

Between November 2012 and July 2013, we deployed camera traps covering the 16 forest ranges of the study area to estimate the status of animal. We divided the study area into 2 km^2 grids and randomly chose grids for camera locations. Within the grid, cameras were predominantly set along park roads; at off-road locations, we installed along game trails and footpaths. Each station consisted of one camera trap of the Moultry D50. All camera traps were programmed to delay sequential photographs by 30 seconds and operate 24 hours per day, recorded time, date and temperature for each exposure. Camera traps were strapped to trees or stakes approximately 50 cm above ground and 1-2 m from the monitoring area. We aimed the censor parallel to the ground to monitor a colonial area approximately 1 m in diametre at 10 m distance. Cameras were checked at 10-14 day intervals for battery replacement and photo download. We aimed to leave camera traps in the forest for the 45 days, but due to work schedule conflicts, cameras were often picked up earlier or later.

Number of trap nights was calculated for each camera location from the time the camera was mounted until the camera was retrieved. After the cameras were retrieved, all photos were downloaded for further study. We identified each photo of an animal to species, recorded the time and date, and rated each photos as a dependent or independent event. Animal detections were considered independent if the time between consecutive photographs of the same species was more than 0.5 hours apart, a convention which follows O'Brien et al. (2003). Because our study was not focused on identifying individuals from photos, so the arbitrary time between independent photos should not introduce bias. Photos with more than one individual of similar species in the frame were counted as one detection for the species.

Camera traps also recorded human traffic (forest staffs, villagers, and poachers), domestic dogs and livestock. Poachers were identified if they were carrying any weapons and animal body parts or ambiguous people visited forest at mid or late night.

Relative abundance of mammals was calculated two ways: RAI. The RAI was calculated for all camera traps mammal species and others based on formula:

$$RAI=\frac{A}{N}\times 100$$

In which 'A' represents the total number of captures of a species by all cameras, and 'N' equal to the total camera traps days during the study period.

The RAI of each species was calculated by to compute all detections for each species are summed for all camera traps over all days, multiplied by 100, and divided by total number of camera trap nights.

Composition of Mammals

We conducted camera surveys at 187 locations, resulting in 6,413 trap days (Mean: 34.48 ± 10.55 SD, range: 9-51). We classified 3,763 frames as independent photographs, of which 6.32% (n =238) were carnivores, 38.4% (n=1446) were non-carnivore mammals, 1.46 (n=55) were of birds, 25.4% (n=955) were villagers, 16.2% (n=611) were staffs, and 8.03% (n=302) were domestic animal. Among these domestic animal 44.37% (n=134) were domestic dogs. We could not determine species in 0.35% (n=13) of the photographs due to poor focus, lighting, or angle. Among the photographs, we identified 24 mammal species (domestic mammals excluded) and seven bird species. The relative abundance of animal is summarized in Table-9.1. The detailed relative abundances of mammal, domestic animal and villagers of each forest range are given in Table-9.1. Among the mammal, two species were endangered, three were vulnerable and three were near threatened species as classified by the 2013 IUCN Red List of threatened species (IUCN, 2013).

Barking deer was the most frequent species (417 detections) followed by the wild boar *Sus scrofa* (290) and hanuman langur (231). Among the globally threatened species, Asian elephant was the most frequent detected species (124 detections) followed by the leopard and sambar. We are also unsure about the precise identification otters. The pictures of the otters looked like the oriental small-clawed otter (*Aonyx cinerea*), which is recently reported in southern part Odisha (Mohapatra et al., 2014). However, because smooth-coated otter (*Lutra perspicillata*) also reported in Similipal and we are not certain of our identification, we categorized simply as otter.

A comparison with a list of large and medium size potential species present in the study area (Annon. 2013) suggests that the completeness of our species recorded was 70.59%. Some of the species were not recorded during our survey (Wild dog *Cuon alpinus* and Four-horned antelope *Tetracerus quadricornis*) may be locally rare as a result of hunting or as a result of widespread presence of human and domestic animals. However, other species like Indian gray wolf *Canis lupus pallipes*, Golden jackal *Canis aureus* and Indian fox *Vulpes bengalensis* have been reported near human habitation of STR. The lack of records of these species may represent a relatively low local abundance. However, these canids relatively prefer open or degraded forests and agricultural areas. The methodology adopted was not species specific, the locations of camera were chosen to cover terrestrial habitat and all the camera traps were installed 50 cm above ground, which would have missed aquatic mammal like otter and the arboreal mammal like Indian giant squirrel *Ratufa indica*. Therefore, the detection probability of otter and Indian giant squirrel might be different between estimated abundance and actual abundance.

Our camera trap data provided the high level of human and domestic animal activities inside the tiger reserve (Fig. 9.1). Human activities usually have important direct (e.g. through hunting) and indirect (e.g. through domestic animals) effects on mammal assemblages. Hunting usually reduces the relative abundance and total biomass of the larger species, occasionally increasing the absolute abundance of the smaller, less preferred ones (Peres, 2010, Peres and Dolman, 2000). Cascades effect through the ecological community can also follow the reduction in the number of the large herbivores (Wright and Duber, 2001), and top predators (Berger et al., 2008). Livestock grazing is an activity that usually has notorious effects on the structure and composition of natural communities (Mathai, 1999, Madhusudan and Mishra,

2003). Among the mammals, large herbivores may be negatively affected by cattle through competitive interactions (Madhusudan, 2004). When prey density is low, the large carnivores usually suffer predation on livestock and villagers see top predators as pests that should be eradicated (Loveridge et al., 2010).

The domestic dogs could also be a problem in the study area, where they accounted for 10.3% detection of anthropogenic photos in camera traps. The abundance and ranging behavior of domestic dogs are recognized as key factors determining their cumulative impacts on wild carnivore through exploitation, apparent and interface competition (Vanak and Gompper, 2010). Dogs were accompanied by villagers and poachers in 48.5% and 8.21% respectively of all dog detections, and the same individual dogs were detected alone. It is possible that some of the dogs detected were feral, and their presence in the study area needs to be address.

Implications for Conservation

Our camera trap data suggest the main threat for wildlife conservation is probably the concomitant increase in incompatible human and domestic animal activities, which currently affect nearly 75% of our study area. To strengthen existing levels of protection in STR, managers need to be made aware of the need to monitor curb threats actively and manage this sensitive ecosystem knowledgably; need to combat poachers with modern approaches through gathering and sharing intelligence, and law enforcement. Many studies suggest that successful conservation results from dedicated protected area management coupled with local community support for the protected area and involvement in its protection (Chauhan et al., 2006; Singh and Gibson, 2011). For Similipal, these actions will be accelerated through involvement of non-government organizations and local communities.

Management strategies for dogs should aim to reduce both the number of dogs and their ranging behavior which determines the spatial extent of their impacts (Vanak and Gompper, 2010; Silva-Rodríguez and Sieving, 2012). Lethal control is a common and effective strategy for population reduction of nuisance predators but is not feasible when such predators are owned, as is the case in several areas where dog impacts have been reported (Lacerda et al., 2009; Silva-Rodríguez et al., 2010; Vanak and Gompper, 2010). This highlights the need to educate people to have fewer dogs, accompanied with reducing the ranging behavior of dogs.

Despite numerous threats, our results suggest that Similipal play an important role in conserving rare and endangered species in this region. This study also provides a framework for further research on biodiversity conservation in this region in the presence of confounding factors. We recommend the need for detailed ecological research and greater awareness among villagers to conserve the biodiversity.

Table-9.1. Relative Abundance Index (RAI) of Wildlife species and others captured photos.

Wildlife	Scientific Name	Food habit	IUCN Status 2013	CT stations	% occurrence in CT stations	Total photos	RAI
Mammals							
Family: Tragulidae							
Indian Spotted Chevrotain	*Moschiola indica*	H	LC	23	12.3	51	0.8
Family: Cervidae							
Barking deer	*Muntiacus muntjac*	H	LC	101	54	417	6.5
Spotted deer	*Axis axis*	H	LC	13	6.95	30	0.47
Sambar	*Rusa unicolor*	H	VU	30	16	89	1.39
Family: Suidae							
Wild boar	*Sus scrofa*	H	LC	99	52.9	290	4.52
Family: Bovidae							
Gaur	*Bos gaurus*	H	VU	3	1.6	4	0.06
Family: Elephantidae							
Asian Elephant	*Elephas maximus*	H	EN	70	37.4	134	2.09
Family: Ursidae							
Sloth bear	*Melursus ursinus*	O	VU	22	11.8	30	0.47
Family: Felidae							
Tiger	*Panthera tigris*	C	EN	1	0.53	1	0.02
Leopard	*Panthera pardus*	C	NT	56	29.9	108	1.68

Wildlife	Scientific Name	Food habit	IUCN Status 2013	CT stations	% occurrence in CT stations	Total photos	RAI
Leopard Cat	*Prionailurus bengalensis*	C		19	2.67	34	0.53
Jungle Cat	*Felis chaus*	C	LC	5	2.67	6	0.09
Family: Hyaenidae							
Striped Hyena	*Hyaena hyaena*	C	NT	1	0.53	2	0.03
Family: Mustelidae							
Otter		C		1	0.53	1	0.02
Ratel	*Mellivora capensis*	C	LC	4	2.14	5	0.08
Family: Viverridae							
Small Indian Civet	*Viverricula indica*	C	LC	11	5.88	14	0.22
Common Palm Civet	*Paradoxurus hermaphroditus*	O	LC	13	6.95	29	0.45
Family: Herpestidae							
Small Indian Mongoose	*Herpestes javanicus*	C	LC	6	3.21	8	0.12
Family: Cercopithecidae							
Rhesus Macaque	*Macaca mulatta*	H	LC	55	29.4	129	2.01
Hanuman Langur	*Semnopithecus entellus*	H	LC	74	39.6	231	3.6
Family: Manidae							
Indian Pangolin	*Manis crassicaudata*	I	NT	1	0.53	1	0.02

Wildlife	Scientific Name	Food habit	IUCN Status 2013	CT stations	% occurrence in CT stations	Total photos	RAI
Family: Leporidae							
Indian Hare	*Lepus nigricollis*	H	LC	9	4.81	19	0.3
Family: Hystricidae							
Indian Crested Porcupine	*Hystrix indica*	H	LC	45	24.1	86	1.34
Family: Sciuridae							
Giant Squirrel	*Ratufa indica*	H	LC	1	0.53	2	0.03
Human and domestic animal							
Forest Department Staff				138	73.8	611	9.53
Department Vehicle				19	10.2	63	0.98
Villagers				140	74.9	955	14.9
Poachers				33	17.6	43	0.67
Livestock				40	21.4	168	2.62
Dogs				66	35.3	134	2.09
Un ID				11	5.88	13	0.2

RAI= Relative abundance index, CT= camera trap, EN= endangered, VU= vulnerable, NT= near threatened, LC= least concern, C= carnivore, H= herbivore, I= insectivore, O= omnivore.

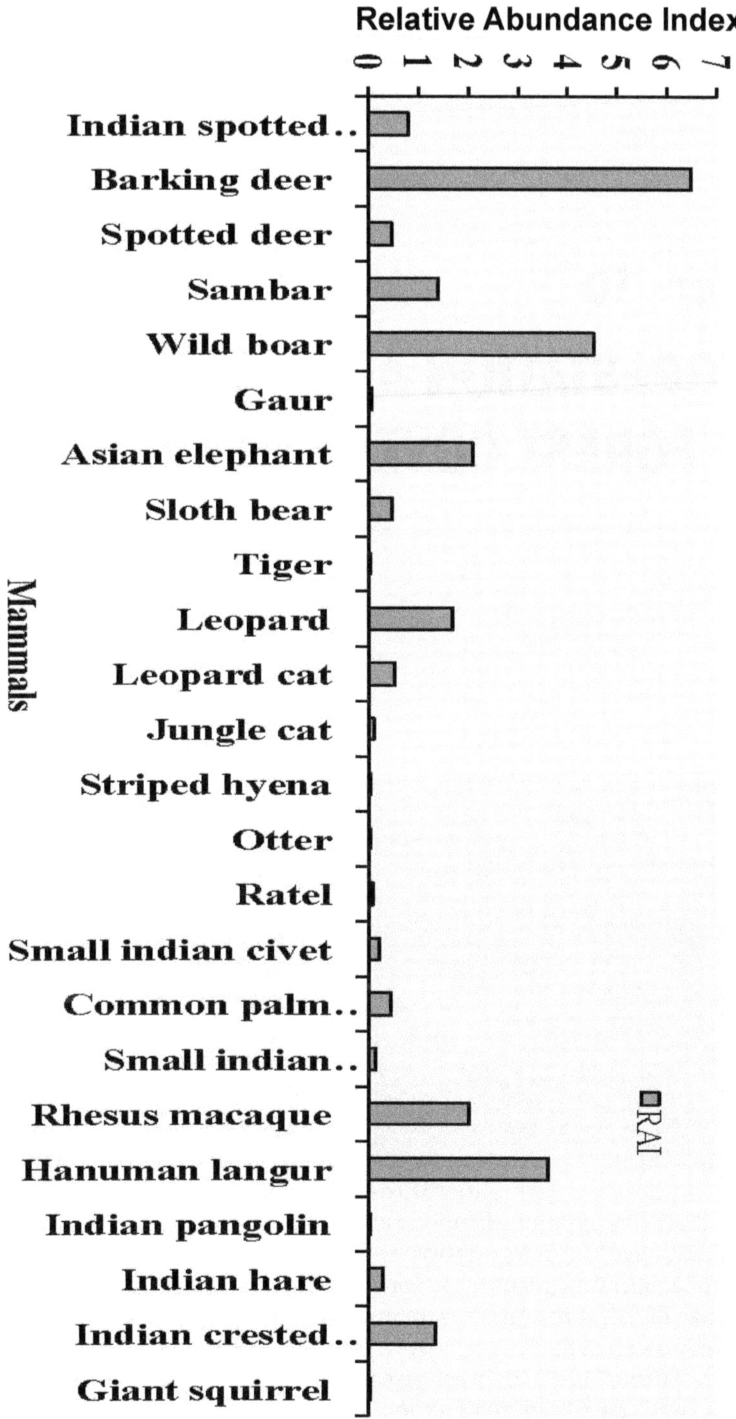

Fig.9.1. Relative Abundance Index of mammals in Similipal Tiger Reserve.

Chapter- 10

DEFORESTATION AND DEGRADATION OF FOREST COVERS IN SIMILIPAL

Deforestation involves a decrease in the area covered by forest. However, degradation involves a reduction of the forest area with a quality decrease in its condition, which is being related to one or a number of different forest ecosystem components (vegetation layer, fauna, soil, etc.), and the interactions between these components, more generally to its functioning. Hundreds of millions of people in developing countries depend on forests for their livelihood. But over 12 million hectares of natural forest are lost in the tropics every year, either through the permanent destruction of forests or through their degradation (Davidar *et al.*, 2007; Geist & Lambin, 2002). One of the important causes of deforestation is the non-sustainable harvesting of forest products for generating income or for sustenance. The factors associated with household use of forest resources is an important starting point for understanding deforestation processes at the local level (Angelsen & Kaimowitz, 1999; Godoy *et al.*, 1997, Spies & Turner, 1999). In most cases, people clear tropical forests to cultivate land. This is motivated by many factors. These include the prospect of generating greater income through farming, changes in land rights, tenure, subsidies, tax laws, resettlement projects, new or restored roads, population pressures and corruption. With respect to Odisha in general and Similipal Biosphere in particular the various factors underlying for degradation and deforestation are mentioned below:

Illegal Cutting of Trees

In Similipal Tiger Reserve, the major source of energy is fuel-wood, and all the households used fuel-wood for their domestic requirements (Davidar *et al.*, 2010). This is because there was no locally available substitute for fuel-wood for the local tribal communities. Fuel-wood is harvested in a daily basis and is therefore an important cause of forest degradation. Most of the upper storey species were harvested by the local people to mitigate their own requirement and to mitigate the fuel wood demand of the locality by supplying it through wood sellers of the nearest market. In the past the people residing on the fringes use to profess on illegal trade of timber and firewood as a supplement to their main income due to bad socio-economic condition and heavy demand of timber and firewood in the growing urban areas surrounding the protected area (PA). In the process, they cut Sissoo (*Dalbergia latifolia*), bija (*Pterocarpus marsupium*), kurum (*Adina cordifolia*), kasi (*Bridelia retusa*) and sal (*Shorea robusta*) trees by axe and saw selectively chronologically in the order and make *in situ* conversion and carry it to outside by head load from where it is transported by bicycles and then by vehicles to the places of destination in the urban areas. Another group of people collect the lops and tops and convert into billets and transport by bicycle for sale as firewood in the local markets. Pressure on forest covers of Similipal with respect to smuggling of economic woods and fire woods throughout the year as evidenced from the amount of timber (in cum.), fire wood (in Mt) and number of poles sized by forest staff on duty. The following figures illustrate the magnitude of illicit felling of trees inside the biosphere reserve during the preceding years (Fig.10.1, 10.2 and 10.3).

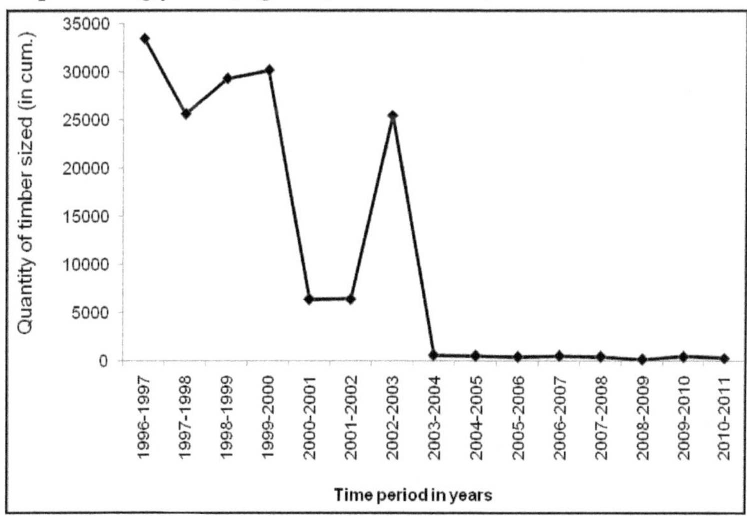

Figure 10.1:Time line data on quantity of timber sized (in cum.) from SBR (Source: STR, Baripada).

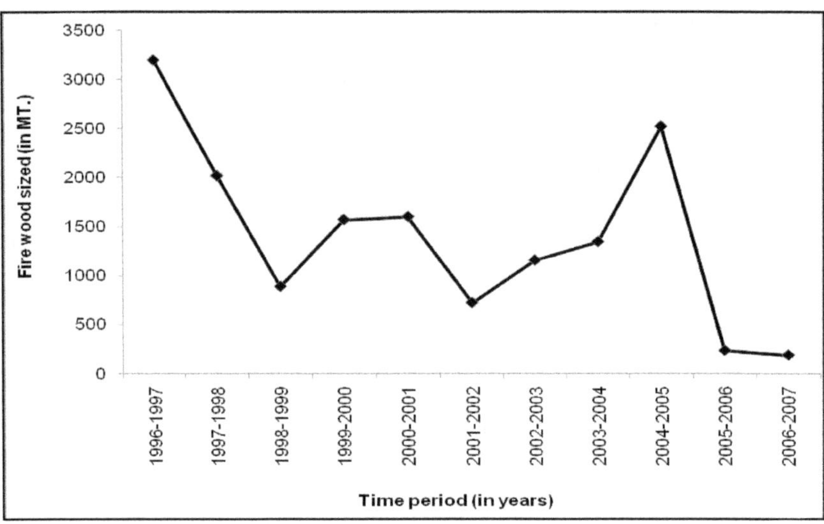

Figure10.2: Time line data on quantity of fire wood sized (in MT.) from SBR (Source: STR, Baripada).

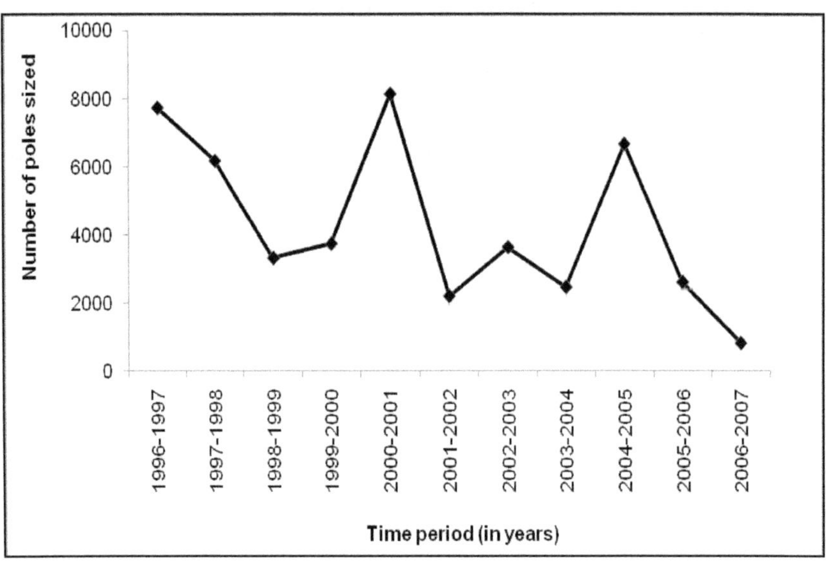

Figure 10.3. Time line data on number of poles sized from SBR (Source: STR, Baripada).

Moreover, Sahoo and Davidar (2013) studied the effect of anthropogenic pressure on plant diversity and vegetation structure of sal forests of Similipal tiger reserve, Odisha and stated that percentage extraction of trees per month was higher in the buffer zone as compared to the core zone (Table-10.1). Variability in extraction pressure in core and buffer areas of the reserve leads to variation in species richness and basal area of the upper storey species.

Table 10.1: Extraction pressure in core and buffer areas of Similipal.

Name of village	Zone	Population	Range of species richness	Range of Basel Area (m^2/ha)	Extraction pressure (% of trees lopped/ month)
Gurguria	Buffer	547	18-23	20.25-33.52	11.06
Nawna	Buffer	216	18-27	22.76-47.14	9.88
Makabadi	Buffer	225	16-22	25.99-45.70	10.40
Pralarampur	Buffer	226	17-25	20.61-47.94	9.56
Bakua	Core	259	22-36	48.62-81.59	1.99
Yamunagard	Core	192	22-37	99.43-112.66	1.77
Kabataghai	Core	187	20-28	56.78-83.52	2.01
Jenabil	Core	260	27-37	108.47-123.34	1.29

Source: Sahoo and Davidar (2013)

Now-a-days illicit felling is spread throughout the year and is more intensive during the rainy season. The timber and firewood on transit are intercepted by Forest staff on duty except some minor offences, which are compounded as per rule, the rest of the offenders are prosecuted in the court of law as evidenced from the number of cases booked with respect to illegal smuggling of timber, fire wood and poles (Table-10.2).

Table 10.2: Number of offence cases booked, cases sent to court and number of offenders arrested.

Year	No. of cases booked	Cases sent to court	No. of offenders arrested
2002-2003	3778	261	520
2003-2004	3174	177	509
2004-2005	5489	237	680
2005-2006	4219	184	618
2006-2007	5705	4227	813
2007-2008	3165	-	172
2008-2009	1729	-	124
2009-2010	3964	-	164
2010-2011	1254	-	71

Source: Office of STR, Baripada

Illegal Removal of NTFP

Harvest of non-timber forest products (NTFPs) has been promoted as an opportunity to both enhance local livelihoods and contribute to biodiversity conservation (Leisher et al., 2010). At the same time, the effectiveness of NTFP harvest at achieving these goals has been criticized (Belcher and Schreckenberg, 2007), and over harvest is considered a major threat to plant diversity (Brummit and Bachman, 2010). Some of the workers stated that unsustainable extraction of forest biomass may cause degradation of forests, loss of biodiversity and ultimately result in the extinction of species due to loss of carrying capacity of the forest and disruption of vital ecological processes (Arjunan *et al.*, 2005; Davidar *et al.*, 2007; Dixon & Sherman, 1990; Karanth *et al.*, 2006; Kothari *et al.*, 1989; Kramer *et al.*, 1992; Madhusudan, 2005; Mammen, 2007; Puyravaud & Garrigues, 2002; Roessingh, 2006; Silori & Mishra, 2001). These two contradictory outcomes illustrates the need to improve the understanding of the circumstances in which NTFP harvest can make a meaningful contribution to local livelihoods without having negative environmental consequences. Rural communities living in fringe villages and inside SBR depends on forests for fuel-wood, fodder and other NTFPs, as a source of cash income or household sustenance. They collect non-timber forest products available inside forest area can be classified under various heads viz. Fibre and flosses, grass, bamboo and canes, edible MFPs, essential oil including those from grasses, medicinal plants, tans and dyes, gums and resins, leaves, fruits and seeds. The use of NTFP items by different tribes of Similipal residing in three different zones, the period of collection, quantity collected and total amount of economy generated from it are mentioned in Table-10.3 and 10.4.

Table 10.3: Seasonality of NTFP's collected by the forest dwellers of Similipal.

Name of NTFP	Months of collection (Collection period)	Tribes collected
Honey	May-June, October	Hill Kharia, Kolha
Wood mushroom	October-February	Hill Kharia, Kolha, Santal
Flower broom	September-April	Hill Kharias
Palua	December-February	Hill Kharia, Kolha
Tasar	August, September	Kolha, Santal, Bathudi
Jhuna	October-December	Kolha, Hill Kharia
Siali leaf	November-March	Bathudi, Kolha, Santal, Hill Kharia
Sal leaf	November-March	Bathudi, Kolha, Santal, Hill Kharia
Mahula flower	April-May	Hill Kharia, Kolha, Santal
Sal seed	June	Kharia and outsiders
Palo Broom	November-January	Munda, Kolha

Contd...

Table 10.3 Contd...

Name of NTFP	Months of collection (Collection period)	Tribes collected
Amla	January	Kharia, Kolha, Santal
Harida	February	Hill Kharia, Bathudi, Santal, Kolha
Bahada	February	Hill Kharia, Kolha, Bathudi, Santal
Bhalia	February	Kolha
Mulika root	Septmber-February	Kolha, Hill Kharia
Sidha fruit	April	Kolha , Santal
Siali fruit	January-February	Kolha , Santal
Kuduchi bark	February-April	Kolha
Dumura	July	Santal, Kolha
Rali fruit	January- December	Hill Kharia, Kolha
Kusuma seed	June-July	Kolha, Bathudi
Kendu	March-April	Kolha, Santal
Chara	March-April	Kolha, Santal, Bathudi
Jamu	June-July	Kolha, Santal, Bathudi
Kendu leaf	March-April	Kolha
Siali bark	October-December	Kolha, Santal
Kochila	September-November	Kolha, Santal
Ambada	January	Kolha, Santal
Tentuli	February	Kolha, Santal, Bathudi
Alu	December-February	Kolha, Santal, Bathudi

Table 10.4: Quantity of NTFP collected and economy generated during 2013-2014.

Name of NTFP item	Unit of collection	Quantity collected	Selling price (in Rs.)
Honey	Kilogram (kg)	39.6	3378
Wood mushroom	Kilogram (kg)	104	1124
Flower broom	Bundles	6.0	927
Palua	Kilogram (kg)	340	6542
Tasar	Piece	83	177
Jhuna	Kilogram (kg)	109	11724
Siali leaf	Bundles	44	3619
Sal leaf	Bundles	90	1953

Contd...

Table 10.4 Contd...

Name of NTFP item	Unit of collection	Quantity collected	Selling price (in Rs.)
Mahula flower	Kilogram (kg)	112	609
Sal seed	Kilogram (kg)	360	1775
Palo Broom	Bundles	46	9375
Amla	Kilogram (kg)	813	4110
Harida	Kilogram (kg)	221	1041
Bahada	Kilogram (kg)	174	876
Bhalia	Kilogram (kg)	21	75
Mulika root	Bundle	382	6253
Sidha fruit	Kilogram (kg)	340	1020
Siali fruit	Kilogram (kg)	67	488
Kuduchi bark	Kilogram (kg)	288	1505
Dumura	Kilogram (kg)	85	850

Studies have shown that in Similipal the tribal forest dwellers and Kharias, a primitive tribe inside the sanctuary subsists on collection of non-wood produce viz. honey, gum, grrowroot and wild mushrooms are collected by them illegally. Besides bark of poja tree (*Litsea monopetala*) is collected by the professional smugglers. Collection of mahua flowers and seeds, sal seeds also cause disturbance to the area. Forests are deliberately burnt to clean the floor to collect these items. However, in the core area the intensity of disturbance is comparatively less than the buffer and trasitional areas. Data collected on year wise collection of NTFP (Pattnaik, 1998) and MFP collected per day (Rout et al., 2010) indicates the pressure exerted on forest covers of Similipal (Table-10.5, 10.6 and 10.7).

Encroachment and Other Illegal Activities

After inclusion of the tract under project Tiger, encroachment has decreased considerably. But clandestine encroachment in the enclaved villages still goes on. Presently there are 1265 villages in SBR. Out of 1265 villages, 61 villages are located in the buffer and 4 villages in core area. The total population of villages located in buffer and core area is 12000 and 449, respectively (Dash and Behera, 2012). In 1200 transitional villages nearly about 0.45 million people were inhabited, out of which 73% are primitive tribes. Out of the 61 buffer villages, 7 are situated very close to the core area boundary. The Government is trying to relocate these 11 villages outside the project area to avoid man-animal conflicts and to reduce future expansion of the villages towards the core area of the biosphere. Human population dynamics of the biosphere is substantially rising inside the biosphere during the last three decades. As per the 1981 census, the total population in the buffer and core of the reserve was 8643 and 314, respectively. By 2001 census, the human population is gradually increased both in the core and buffer area of the reserve and reached up to 12000 and 449, respectively, implying that the population has increased by 40% (Table-10.8). The increased population imposed pressure on the density and diversity of plants and animals of the biosphere to fulfil their day to day needs and in the future they may encroaching the forestlands for their habitation.

Table- 10.5. Year and item wise collection of NTFP from Similipal

Year	1979-80	1980-81	1981-82	1982-83	1983-84	1984-85	1985-86
Sal Seeds	0.85	31.72	5552.55	17014.35	2892.53	9444.07	12576.56
Mahua Seeds	24.12	7.85	149	0.3	48.14	67.32	1.4
Kusum Seeds	2.27	0.1	311.55	1.28	35.36	0.33	20.09
Karanj Seeds	0.18	1.18	40.85	2.99	0.53	2.46	0.6
Char Seeds		1.23		0.85	0.49		
Mango Kernel	13						
Honey	14.17	11.61	15.8	10.49	3.35	3.63	11.82
Wax	0.25	0.15	1.35	0.08	0.12	0.01	0.06
Resin	0.4	1.7	1.01	5.75	32.65	15.8	4.46
Arrowroot	0.12	1.15	1.61	2.02	8.11	0.44	1.22
Myrobalans		22.52	2.88	0.05			5.8
Aonla		7.38	0.04				
Siali fibre		0.24	0.69				
Tamarind		1.39	352.41	388.14	9.65	70.35	104.27
Genduli gums		0.14	0.41				
Antha gums		1.3	20.86	91.33	0.66	5.86	8.59
Nux vomica			0.22	8.7			
Nageswar			1.32				
Wild citrus			0.05				
Semul cotton						0.15	

Contd...

Table 10.3 Contd...

Year	1986-87	1987-88	1989-90	1990-91
Sal Seeds	11361.81	3203.21	25379.45	274.15
Mahua Seeds			68	
Kusum Seeds	0.5	0.13		
Karanj Seeds				
Char Seeds				
Mango Kernel				
Honey	8.08	8.08	6.55	7.89
Wax	0.04	0.04		0.03
Resin	14.45	40.63	7.15	0.845
Arrowroot	0.65	1.49		
Myrobalans	19.24	29.23	0.3	
Aonla				
Siali fibre				
Tamarind	36.44	0.45		
Genduli gums				
Antha gums	4.24	15.45	7.04	
Nux vomica				
Nageswar				
Wild citrus				
Semul cotton	8.05			

Source: Pattnaik (1998)

Table-10.6. Year wise collection of MFP items from Similipal.

Year	Name of the M.F.P. Items	Quantity Collected
1990-91	Honey	1.80 Qntls
	Arrowroot	0.850 Qntls
1991-92	Honey	13.320 Qntls
	Resin	32.690 Qntls
	Arrowroot	9.750 Qntls
1992-93	Honey	134.826 Qntls
	Arrowroot	11.740 Qntls
1993-94	Honey	203.445 Qntls
	Arrowroot	2.000 Qntls
1994-95	Honey	10.470 Qntls
	Arrowroot	0.570 Qntls
1995-96	Honey	12.630 Qntls

Source: Pattnaik (1998)

Table-10.7. Quantity collection of MFP items by individuals in Similipal.

Name of The MFP item	Quantity collected per Day
Honey	2kg-10kg
Resin	5kg-8kg
Arrowroot	1/2kg-5kg
Wax	1kg
Kusum Seeds	10kg-20kg
Karanj Seeds	15kg-25kg
Gum	5kg-8kg
Sal Seeds	30kg-60kg

Source: Rout et al. (2010)

Table-10.8: Demography of core and buffer villages of Similipal Biosphere Reserve (SBR).

Name of village	Zone	Name of block	Name of GP	Population as per 2001 census			
				SC	ST	Others	Total
Bakua	Core	Jashipur	Astakumar	0	80	0	80
Jamunagarh	Core	Jashipur	Astakumar	0	96	0	96
Kabatghai	Core	Jashipur	Astakumar	0	115	0	115

Contd...

Table 10.8 Contd...

Name of village	Zone	Name of block	Name of GP	Population as per 2001 census			
Jenabil	Core			-	-		260
Asanbani	Buffer	Jashipur	Astakumar	19	240	0	259
Bad-kasira	Buffer	Jashipur	Astakumar	0	222	8	230
Bad-uski	Buffer	Jashipur	Astakumar	0	232	9	241
Barigaon	Buffer	Jashipur	Astakumar	0	150	32	182
Bharrachua	Buffer	Jashipur	Astakumar	0	90	0	90
Bilapagha	Buffer	Jashipur	Astakumar	4	257	6	267
Chandikhaman	Buffer	Jashipur	Astakumar	0	156	2	158
Gudgudia	Buffer	Jashipur	Astakumar	9	374	143	526
Khariadunguri	Buffer	Jashipur	Astakumar	0	110	0	110
Khejuri	Buffer	Jashipur	Astakumar	0	366	52	418
Kohla	Buffer	Jashipur	Barehipani	0	220	7	227
Kondabil	Buffer	Jashipur	Barehipani	0	342	30	372
Kuanribil	Buffer	Jashipur	Barehipani	0	344	2	346
Kumari (kumbhari)	Buffer	Jashipur	Barehipani	1	254	38	293
Kusumi	Buffer	Jashipur	Barehipani	10	112	109	231
Nenjaghosra	Buffer	Jashipur	Barehipani	0	84	1	85
Nuniaguda	Buffer	Jashipur	Barehipani	0	182	0	182
Routola	Buffer	Jashipur	Barehipani	0	246	13	259
Sankasira	Buffer	Jashipur	Barehipani	0	184	0	184
San-uski	Buffer	Jashipur	Barehipani	0	233	0	233
Saharpat	Buffer	Jashipur	Barehipani	1	236	61	298
Astakumar	Buffer	Jashipur	Barehipani	0	351	94	445
Balarampur	Buffer	Jashipur	Barehipani	0	318	6	324
Bandribasa	Buffer	Jashipur	Ektali	0	240	0	240
Barheipani	Buffer	Jashipur	Gurguria	0	445	28	473
Barsia	Buffer	Jashipur	Gurguria	0	399	4	403
Burhabalanga	Buffer	Jashipur	Gurguria	0	249	0	249
Chakundakocha	Buffer	Jashipur	Gurguria	0	29	0	29
Garh-similipal	Buffer	Jashipur	Gurguria	0	333	84	417
Gopinathpur	Buffer	Jashipur	Gurguria	0	217	0	217
Haladia	Buffer	Jashipur	Gurguria	0	132	0	132
Jadadihi	Buffer	Jashipur	Gurguria	0	126	0	126
Kiajhari	Buffer	Jashipur	Gurguria	0	154	31	185

Table 10.8 Contd...

Name of village	Zone	Name of block	Name of GP	Population as per 2001 census			
Kukurbhuka	Buffer	Jashipur	Gurguria	0	300	0	300
Kuljhari	Buffer	Jashipur	Gurguria	0	78	0	78
Lembujharan	Buffer	Jashipur	Gurguria	0	123	0	123
Makabadi	Buffer	Jashipur	Gurguria	0	228	13	241
Nawana	Buffer	Jashipur	Gurguria	0	292	13	305
Nikhirda	Buffer	Jashipur	Gurguria	0	58	2	60
Phulbari	Buffer	Jashipur	Gurguria	0	80	0	80
Rajupal	Buffer	Jashipur	Gurguria	0	148	0	148
Saruda	Buffer	Jashipur	Podagada	0	139	0	139
Charbandha	Buffer	Bangiriposi	Brahman gaon	0	449	0	449
Dantiakocha	Buffer	Bangiriposi	Brahman gaon	0	89	0	89
Alapani	Buffer	Bangiriposi	Sarispal	0	113	0	113
Amdapani	Buffer	Bangiriposi	Sarispal	0	170	12	182
Barubeda	Buffer	Bangiriposi	Sarispal	0	162	0	162
Basilakacha	Buffer	Bangiriposi	Sarispal	0	57	0	57
Bhoduakocha	Buffer	Bangiriposi	Sarispal	0	23	0	23
Chakidi (chakidih ipanipal)	Buffer	Bangiriposi	Sarispal	0	155	0	155
Jamtolia	Buffer	Bangiriposi	Sarispal	0	107	0	107
Jerkani	Buffer	Bangiriposi	Sarispal	0	88	0	88
Khadighati	Buffer	Bangiriposi	Sarispal	0	78	0	78
Kukurbhuka	Buffer	Bangiriposi	Sarispal	0	147	0	147
Kusumtota	Buffer	Bangiriposi	Sarispal	0	65	0	65
Phuljhari	Buffer	Bangiriposi	Sarispal	0	138	0	138
Purunapani	Buffer	Bangiriposi	Sarispal	28	213	1	242

Source: Office of The STR, Baripada

At present a new problem in the shape of insurgent activities has crept in. As the PA is situated very close to Jharkhand and West Bengal, the insurgent have found it to be a safe asylum for them since there is no road communication to the place. They gain the confidence of the local people inside the PA by instigating them against the action of field functionaries of the Sanctuary. By gaining confidence of the local community inhabited inside the biosphere reserve, the immigrants coming from Jharkhand and West Bengal area are engaged in smuggling of timber and export of body parts of wild animals to outside through poaching of wild animals. In the transition zone between forests and villages identification of girdled trees manifest this fact. Detailed surveys need to be taken up to ascertain the exact nature of problem.

Human Impact on Similipal Forests

Similipal forests have been suffering heavy damage over the past centuries and this beautiful region is currently under severe threat from human pressure (Upadhyay et al., 2012). Human disturbance has altered the vegetation and biodiversity. Large-scale agriculture and development, and extraction of forest products in the form of grazing, lopping, litter removal and NTFPs for commercial use, has extensively degraded the forests (Parida, 1997; Raut & Behera, 1997). The intense use of forests has retarded regeneration and caused changes in the abundance and spatial distribution of plants (Rath & Sutar, 2004). This unique natural forest is now under threat due to a number of factors and the major cause is the unsolved socio-economic issues. Human impact on Similipal forests needs to be measured and management solutions provided so that degradation of the forests can be arrested.

There were 61 villages located inside STR and about 1200 villages along the periphery with more than 0.45 million people who depended on the forests to meet their day to day requirements and supplemented their income through collection of NTFPs. The population growth rate among three gram panchayats ranged from 1.8 to 2.7% among the non-tribal population and 1.8 to 2.3% among the tribal population (Census of India 1991, 2001). Agriculture was the main occupation supplemented by livestock raring, NTFP collection and annual Akhand Shikar (annual mass hunting) (Rath & Sutar, 2004). As evidenced with respect to land uses of inhabitants of SBR a study was carried out by Upadhaya et al. (2012) taking some representative core and buffer villages of SBR and they pointed out in their study about the types of land uses. They mentioned in their study the degraded land and culturable waste land is more in the buffer villages than the core village. They also observed in their study the ratio of culturable waste land to agricultural land and degraded land to agricultural land is high in the buffere villages than the core villages of the reserve (Table-10.9).

Considering the study conducted by Upadhyaya et al. (2012) as a case study the level of human pressure on forest lands and conversion of forest lands for agricultural activity and other purposes is clearly understandable. However, more and more systematic and scientific data of human impact on forest lands of Similipal is required to take proper management decisions and to escape the forest covers of SBR from human pressure for future sustainability.

Domestic Livestock Grazing

Grazing is prohibited in the core and buffer areas of the Similipal sanctuary. But incidence of grazing is noticed around the villages of core area as well as the other villages located just adjacent to core area. There are 65 villages inside PA. The people rear cattle for the purpose of cultivation of the crop only. They do not mulch the cow. Hence, unlike other areas, they retain the cattle during 4 months a year coinciding with cultivation and for rest 8 months they release them into the forest. In the fringe areas, the people graze the cattle and collect fodder from the forest. The movement and grazing of cattle have restricted the movement of wild animals in the buffer area and so also inside the sanctuary around the villagers due shortage of forages. The impact of grazing can be well imagined from the grazing animal population of the villages (Table-10.10).

Table-10.9: Details of different types of land uses of some representative core and buffer villages of SBR.

Types of land use	Core village	Buffer villages	
	Jenabil	Nawna	Ghodabindha
Forest land (ha)	35.24	231.59	118
Agriculturable land (ha)	74.61	98.98	155
Culturable waste land (ha)	2.0	62.57	39
Degraded land (ha)	27.45	146.60	80.56
Agriculture: Forest land	2.12	0.42	1.31
Culturable waste land : Agricultural Land	0.03	0.63	0.25
Degraded land: Agricultural land	0.37	1.48	0.52

Source: Upadhyay et al. (2012)

Wild Fire

Fires are common and frequent due to high deposition of combustible materials like dried grass, a mat of leaf litter and other debris like fallen parts of trees. Barring certain moist valleys, entire Similipal hills comes under conflagration although fires generally confined to ground only. The causes of fire are purely anthropogenic mainly by MFP collectors, hunters, poachers and callous tourists. Fire season commences in mid February and continues up to mid June normally. However, a sporadic rain in pre-monsoon period and summer storms curtails the intensity of fires. The main reasons of forest fire in Similipal area are:

(a) For collection of mohua flowers, sal seeds and other items of minor forest produce both in plain forests and hills, the villagers set fire to the dry leaves along with ground flora to facilitate the collection of above articles easily.

(b) In Similipal hills there is an abundance of large variety of grasses. The density of grasses varies from being then in densly wooded areas to dense in open hill sides and hill tops where the forest appears a typical savannah appearance. Besides these, there are abandoned cultivated areas which are full of dense grasses. These grasses make the Similipal hills inflammable during the fire season.

(c) The professional graziers also set fire to the forest in the early part of the summer season to get new succulent grass for grazing.

(d) The Kharias who live inside the Similipal hills set fire to the forest for collection of honey and wax, shooting, trapping and hunting animals.

(e) The timber contractors who camp inside their lease areas for conversion and extraction of timber also some times cause forest fires largely on account of their negligence.

(f) During the Akhand shikar in the month of April which is part of the social custom of the tribal population there is large-scale setting of fires in the forests to facilitate communal hunting.

(g) The forest fire is also often caused by the nearby villagers, travelers, picnic parties, etc. by negligence.

Table-10.10: Grazing animal population of SBR.

Name of villages/ Core/Buffer	Cattle	Buffaloes	Goats	Sheep	Source of data
Kabat Ghai	235	18	138	-	Office of STR
Jamunagarh	130	01	136	23	Office of STR
Jenabil	262	-	188	01	Office of STR
Bakua	156	-	78	-	Office of STR
Gurguria	131	-	38	-	Upadhyaya et al. (2012)
Core (total:4 villages)	783	-	-	-	Office of STR
Buffer (total: 57 villages)	17016	-	-	-	Office of STR
Core (average)	195.75	-	-	-	Office of STR
Buffer (average)	298.53	-	-	-	Office of STR

The total area affected under forest fire was estimated as: 860.9 km^2 in 2004, 418 km^2 in 2005, 902.5 km^2 in 2006, 855.3 km^2 in 2007 and 653.6 km^2 in 2008, 1014.7 km^2 in 2009, 594.9 km^2 in 2010, 592.8 km^2 in 2011, 982.2 km^2 in 2012 and 508.2 km^2 in 2013 (Fig. 9). The annual burnt area proportionately is in the minimum range of 11.5% in 2005 to maximum of 27.8% in 2009 of total vegetation area. During the last decade, year 2009 have more anomalous amount of fire incidences (IMD, 2009). The mean annual rainfall was 600 mm during 2009. Average estimate based on decadal data reveals that every year an area of 738 km^2 (20.2%) is experiencing forest fire.

The previous study by Odisha Forest Department has recorded forest burnt area as 221 km^2 in 1995, 170.5 km^2 in 1996 and in 103.9 km^2 in 1997. It indicates increasing trend of forest fires during the recent period. In temporal scale the burnt area forest patches (km^2) during 2004 to 2013 is represent in Fig. 10.4.

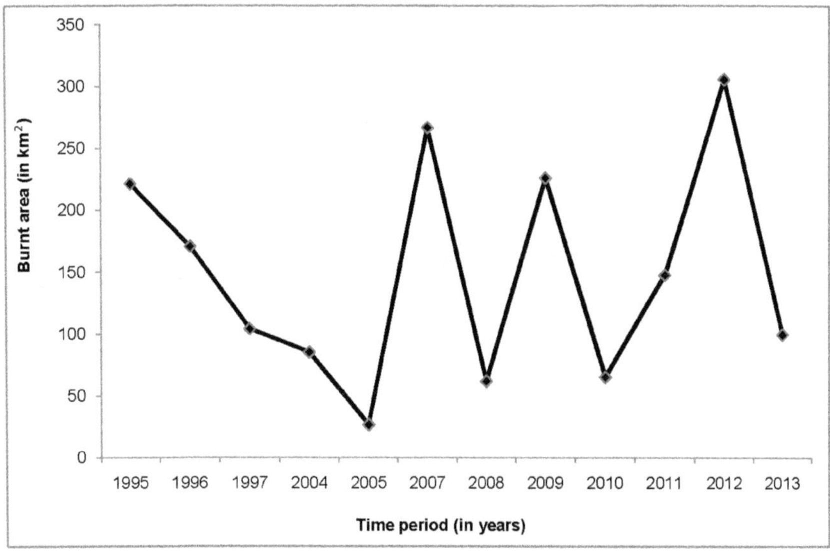

Fig.10.4. Year wise burnt area (in km^2) of Similipal biosphere reserve. (Source: Saranya et al., 2014)

Fire incidence of different vegetation types of Similipal at different time periods is presented in Fig.10.5 Saranya et al. (2014) observed that forest fires were detected in nine vegetation types. However, they excluded the riparian forest from their burnt area analysis, due to its proximity to water channels and extremely moist condition throughout the year. The proportion of total burnt area among the individual vegetation types varies considerably. Semi-evergreen forests mostly remain protected from fires due to high content of soil moisture and predominance of evergreen tree elements. A low density of burnt area patches were detected in the borders to the inner part of semi-evergreen forests which are located at the edge of the grasslands, savannah and deciduous forests. A relatively low percentage of burnt area affected the semi-evergreen forests in the form of small and scattered patches of burnt areas. Of the nine vegetation types, moist deciduous forest shows significantly high burnt area coverage. But fire extent in the dry deciduous and savannah was significantly higher than the moist deciduous type. It is due to dry inflammable material in the form of grass and leaf litter was significantly higher in dry deciduous forests and savannah. Prolonged dry weather conditions, also seems to be responsible for spread of more fires. Forest fires were very severe in all vegetation types during 2009, which was the warmest year since 1901 (IMD, 2009). The high level Sal forests are subjected to annual fires and frost. As a result the forests tend to become open and in extreme cases a steady retrogression to grassland takes place (Saxena & Brahmam, 1989). Decadal analysis reveals dry deciduous forests, tree savannah, shrub savannah; grasslands and high level Sal forests are mostly affected by forest fires (Fig.10.5).

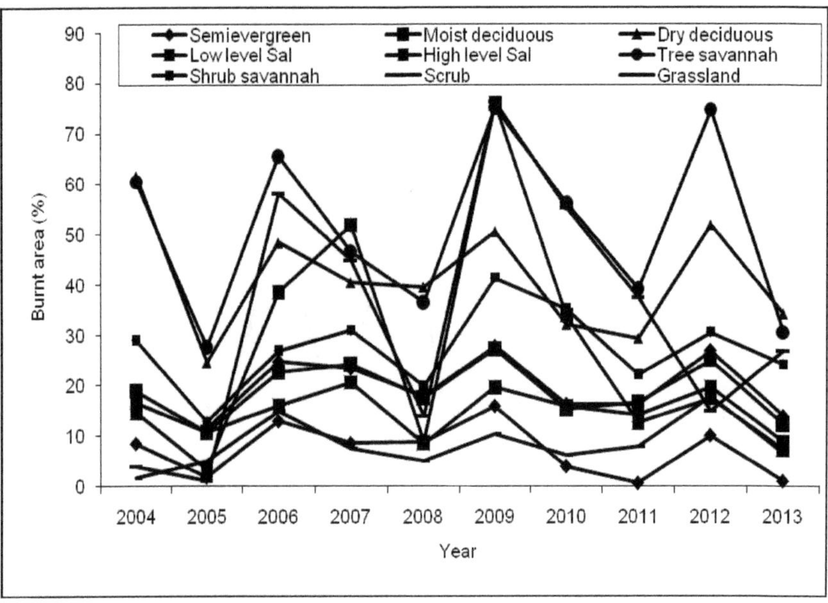

Fig.10.5 Year wise burnt area (in percentage) of different vegetation types of Similipal biosphere reserve. (Source: Saranya et al., 2014)

Insect Attacks and Pathological Problems

These aspects are not properly documented as far as the intensity of their damage is concerned. The trees like *Madhuca latifolia, Buchanania lanzan* and *Terminalia alata* are occasionally infested by *Dendrophthoe falcate*. During May-July a green caterpillar defoliates the sal tree. However, the damage caused is not so serious and alarming. Grasshoppers occur in huge swarms during rainy season but their damage is not so acute. A healthy population of birds may be the reason for low insect damage, A detailed entomological; and pathological investigation is quite essential to ascertain the real status of these two groups of organisms.

Soil Nutrient Status and Degradation

Nutrients of soil play an important role in plant growth. It is directly involved in nutrition of plants and role played by one element cannot be replaced by any other. Nutrition is a factor of paramount importance that regulates growth, development and reproduction of plants. Intake and growth targets are important to reach the functional optima. Nitrogen makes the plant green, increases the protein content and encourages good quality foliage. It is also an important constituent of chlorophyll. Most of the soil nitrogen occurs in the organic combination. Decomposition of the organic carbon by breakdown of protein provides soil nitrogen in ionic form (Bremner, 1951). Phosphorus stimulates the root formation makes it drought resistant and increases the protein content in the leaf. It affects rapid growth. Potassium helps in increasing nitrogen, helps in the building of protein, photosynthesis and reduces diseases in plants. Loss of surface layer of the soil due to anthropogenic activity leads to decrease in soil nutrient status which ultimately affects the growth and development of natural vegetation in natural forest covers. Tripathi and Singh (2012) studied the Physico-chemical characterization of soils of forest and savanna ecosystems of Similipal at different depths and seasons (Table-10.11). They observed that mean values across the forest types and depths indicated significant differences due to site and season. Savannization caused decline in nitrate N in both soil depths by 21% and 23%, and increased in ammonium N by 37% and 46%, respectively. The conversion of forest into savanna at both soil depths (0-10cm and 10-20cm), increased significantly the bulk density by 18.33 and 13.15%, and decreased the OC, TN and TP by 40 and 18%, 38 and 30%, and, 13 and 39%, respectively (Table-10.11).

Table- 10.11: Physico-chemical characterization of soils of forest and savanna ecosytems (means ±1 S.E.) at different depths.

Parametres		Sal forest		Mixed forest		Savanna (*Imperata*)		Savanna (*Heteropogon*)	
		0-10	10-20	0-10	10-20	0-10	10-20	0-10	10-20
Soil particle size	>0.2mm (%)	66 ± 0.7	60 ±0.7	68 ± 0.5	60 ±0.7	72± 1.0	67 ±0.7	75 ±1.4	67 ±0.7
	0.2-0.1mm	18 ±0.4	19 ±0.4	20± 0.5	20.2 ±0.015	19 ±0.4	23 ±0.067	14 ±0.36	21± 0.42
	<0.1mm	16 ±0.8	21± 0.7	13± 0.8	17 ±0.7	8 ±0.7	10 ±0.7	10 ±1.2	12± 0.93
Bulk density(g/cm^3)		0.97 ±0.17	1.12 ±0.028	1.08± 0.013	1.16 ±0.033	1.20 ±0.012	1.28 ±0.021	1.22 ±0.013	1.30 ±0.017
WHC (%)		38.96 ±0.052	35 ±0.052	36.41± 0.032	34 ±0.04	32.98 ±0.05	29 ±0.12	34.33± 0.034	31 ±0.096
Moisture (%)		10.32 ±0.02	8.62 ±0.02	9.17 ±0.017	8.10 ±0.02	7.72 ±0.12	6.35± 0.02	7.41 ±0.12	6.10 ±0.02
Temp (^{0}C)		20.2	18	22.5	20.5	24.40	22.2	24.6	22.5
pH		5.76 ±0.02	5.22 ±0.22	6.01 ±0.02	5.68 ±0.02	4.78 ±0.011	4.55 ±0.02	4.58 ±0.011	4.25 ±0.02
Organic Carbon (%)		1.98 ±0.02	0.8 ±0.16	1.84 ±0.026	0.75 ±0.017	1.11 ±0.02	0.62 ±0.018	1.17 ±0.02	0.65 ±0.022
Total-N (%)		0.362 ±0.02	0.095 ±0.02	0.346± 0.019	0.085 ±0.033	0.21 ±0.02	0.062 ±0.034	0.23 ±0.017	0.065 ±0.018
Total-P (%)		0.045 ±0.003	0.014 ±0.027	0.039 ±0.004	0.014± 0.006	0.038 ±0.003	0.0135 ±0.002	0.035 ±0.005	0.0135 ±0.003

Source: Tripathi and Singh (2012)

Extensive studies were undertaken in Simlipal biosphere reserve, on soil chemical changes that occur due to deforestation and cultivation. The study revealed that deforestation and cultivation result in significant reduction in organic carbon, total nitrogen, C:N ratios, cation exchange capacity, available calcium and magnesium. However, there are no significant changes in total and available phosphorus, available potassium and sodium; C:P and N:P ratios. Loss of organic carbon and nitrogen occurs rapidly in the first 15 years of cultivation and reaches quasi-steady state values around 1-2% organic carbon and 0.1-0.2% nitrogen. Extent of reduction is not related to initial levels. Evergreen forest soils have the highest levels of organic carbon and nitrogen, followed by deciduous forests, grasslands and cultivated lands in that order. Cation exchange capacity, total exchangeable bases, exchangeable calcium and potassium levels are also highest in evergreen forest soils; deciduous forests, grasslands and cultivated soils have statistically similar contents of bases. Exchangeable magnesium is not affected by vegetative cover. Cultivation causes some reduction in acidity but there is no reduction in clay content of the soil. Cultivated soils contain much less moisture than the adjoining forest soils and evergreen forest soils have the highest moisture contents. It is concluded that soil organic matter, moisture and cation exchange capacity are the key factors of soil exhaustion. The removal of green forest canopy results in a cessation of litter input and drying of the soil. Thereby, conditions for optimum turnover of humus are no longer present, organic carbon and nitrogen is lost to the atmosphere, soil exchange capacity is lowered with loss of calcium and magnesium.

The Illegal and Unscientific Collection of Medicinal Plants

Similipal Biosphere Reserve (SBR) is home to approximately 1076 plant species are used in various purposes. Out of 1076 plant species 600 species representing 74 families in 34 compartments are used in the treatment of different diseases. Tribals living inside and outside the forest cover of SBR possess knowledge about these medicinal plants and their utilization in treatment against various diseases. Various ethnomedicinal plants and their parts has been documented against the diseases like gastrointestinal disorder, skin diseases, gynecological disorder, skeletal diseases, jaundice, piles, bronchitis, diabetes, neurological diseases, snake bite, ophthalmic infection and cardiovascular diseases, etc. The illegal and unscientific collection by local people, and over exploitation by village vaidyas and local kavirajs has led to the drastic reduction in the number of medicinal plant species like *Asparagus racemosus, Boerhavia diffusa, Dioscorea* sp., *Helicteres isora, Litsea monopetala, Rauvolfia serpentina, Saraca ashoka* and *Smilax macrophylla*, etc. Since the exploitation of these valuable resources is increasing rapidly, many species of medicinal plants are becoming rare and are included in the list of endangered plants.

Poaching of Faunal Species

Year wise sporadic cases of organized poaching of tiger, leopard, bear, elephant, samber, deer, barking deer, spotted deer, wild boar, gurandi, wild pig, giant squirrel, langur, peacock, wild cat, etc. are documented by the Office of the Field Director, Similipal Tiger Reserve, Baripada. The death abstract of wild animals in SBR from time to time is mentioned in Table-10.12. The death record of wild animals in Similipal indicates about the threat imposed on them by the forest dwellers of Similipal residing in different zones.

Table-10.12. Year wise death record of wild animals of Similipal.

Species	**2005-06**	**2006-07**	**2007-08**	**2008-09**	**2009-10**	**2010-11**	**2011-12**	**Total**
Tiger	-	-	-	-	-	1	-	1
Leopard	-	-	-	3	-	-	-	3
Bear	-	-	-	2	-	-	-	2
Elephant	2	3	4	4	1	14	9	37
Sambar	3	4	-	2	3	5	2	19
Deer	1	1	-	-	-	-	1	3
Barking deer	1	7	1	1	4	1	-	15
Spotted deer	-	-	-	-	1	-	-	1
Wild boar	1	4	4	-	-	1	-	10
Gurandi	-	-	-	1	2	-	-	3
Wild pig	-	-	-	1	-	-	-	1
Giant squirrel	-	1	-	-	-	-	-	1
Langur	-	-	-	-	1	-	-	1
Peacock	-	1	-	-	-	-	-	1
Wild cat	-	-	-	-	-	2	-	2
Total	8	21	9	14	12	24	12	**100**

REFERENCES

Aide, T.M. (1988). Herbivory as selective agent on the timing of leaf production in a tropical understorey community. *Nature*, 336: 574-575.

Airi, S., Rawal, R. S., Dhar, U. and Purohit, A. N. (2000). Assessment of availability and habitat preference of Jatamasi a critically endangered medicinal plant of west Himalayas. *Current Science*, 79: 1467-1470.

Alder, D. and Synnott, T. J. (1992). Permanent Sample Plot Techniques for Mixed Tropical Forest. *Tropical Forestry Papers* 25, University of Oxford, UK.

Anderson, J. M. and Ingram, J. S. I. (1993). Tropical Soil Biology and Fertility: A Handbook of Methods. 2nd Edn. CAB International, Wallingford, UK.

Arroyo, M.T.K. (1979). Comments on breeding system in Neotropical forest. In: Larsen, K. (Ed.). *Tropical Botany*. Academic Press ,London.

Anderson, J. M. and Swift, M. J. (1983). Decomposition in tropical forests. In Sutton, S. L., Chadwick, A. C. and Whitmore, T. C. (Eds.). *Tropical Rain Forest: Ecology and Management*. Blackwell Scientific Publications, Oxford.

Andrea, C. A., and Russell, G. (2005). Are epiphytes important for birds in coffee plantations? an experimental assessment. *Journal of Applied Ecology*, 42: 150-159.

Angelsen, A. and Kaimowitz, D. (1999). Rethinking the causes of deforestation: Lessons from economic models. *The World Bank Research Observer*, 14: 73–98.

Anonymous (2012). Annual Report. Similipal Tiger Reserve, Odisha, India.

Arjunan, M., Puyravaud J. P. and Davidar, P. (2005). The impact of resource collection by local communities on the dry forests of the Kalakad–Mundanthurai Tiger Reserve. *Tropical Ecology*, 46 :135–144.

Arjunan, M.C. and Pannammal, N. R. (1993). Studies on phenology and nursery technology of certain tree species. *Journal Indian Botanical Society*, 10: 147-150.

Barbour, M. G., Burk, J. H. and Pitts, W. D. (1980). Terrestrial Plant Ecology. The Benjamin /Cummings Publishing Company, Inc., Menlo Park, CA.

Berg, B. and Staaf, H. (1981). Leaching, accumulation and release of nitrogen in decomposing forest litter. In: Clark, F. E. and Rosswall, T. (Eds.). Terrestrial Nitrogen Cycles: Processes, Ecosystem Strategies and Management Impacts. O¨sterfa¨rnebo, Sweden.

Biswal, A. K., Nair, M. V. and Upadhyay, H. S. (2011). Flora and Fauna of Similipal Biosphere Reserve. Vol. I, Similipal Tiger Reserve, Mayurbhanj, Baripada.

Boonyawat, S. and Ngampongsai, C. (1974). An Analysis of Accumulation and Decomposition of Litter Fall in Hill Evergreen Forest, Doi Pui, Chiangmai. Kogma Watershed. Research Bulletin, Kasetsart Universtiy Press, Thailand.

Brinson, M. M. (1977). Decomposition and nutrient exchange of litter in an alluvial swamp forest. *Ecology*, 58: 601–609.

Bunnell, F. and Tait, D. (1974). Mathematical simulation models of decomposition processes. In: Holding, A. J. (Ed.). Soil Organisms and Decomposition in Tundra. Tundra Biome Steering Committee, Stockholm, Sweden.

Burtt Davy, J. (1938). The classification of tropical woody vegetation types. *Imp. For. Inst.,* Paper No. 13.

Babweteera, F., Plumptre, A. and Obua, J. (2001). Effect of gap size and age on climber abundance and diversity in Budongo forest reserve, Uganda. *African Journal of Ecology,* 38: 230-237.

Barritt, A. R. and Facelli, J. M. (2001). Effects of *Casuarina pauper* litter and grove soil on emergence and growth of understorey species in arid lands of South Australia. *Journal of Arid Environments*, 49: 569-579.

Behangana, M. (2004). The diversity and status of amphibians and reptiles in the Kyoga lake basin. *African Journal Ecology*, 42: 51-56.

Belcher, B. and Schreckenberg, K. (2007). Commercialisation of non-timber forest products: a reality check. *Dev. Policy Rev.*, 25:355-377.

Benkobi, L., Trilica, M. J. and Smith, J. L. (1993). Soil loss as affected by different combinations of surface litter and rock. *Journal of Environmental Quality*. 22: 657-661.

Berger, K.M., Gese, E.M., Berger, J. (2008). Indirect effects and traditional trophic cascades: a test involving wolves, coyotes and pronghorn. *Ecology*, 89:818-828.

Bhat, D. M. and Murali, K. S. (2001). Phenology of understorey species of tropical moist forest of Western Ghats region of Uttara Kannada district in south India. *Current Science*, 81(7): 799-805.

Bhat, D.M., Naik, M.B., Patagar, S.G., Hedge, G.T., Kanade, Y.G., Hedge, G.N., Shastri, C.M., Shetti, D.M. and Furtado, R.M. (2000). Forest dynamics in tropical rain forests of UttaraKannada district in western ghats, India. *Current Science*, 79:975–985.

Boojh, R. and Ramakrishnan, P.S. (1981). Phenology of trees in a subtropical evergreen montane forest in northeast India. *Geo-Eco-Trop.*, 5: 189-209.

Borchert, R. (1994). Soil and stem water storage determine phenology and distribution of tropica dry forest trees. *Ecology*, 75: 1437-1449.

Bremner, J.M. (1951). A review of recent work on soil Organic matter. *J. Soil Sci.*, 2: 67-82.

Brummit, N. and Bachman, S. (2010). Plants under pressure: A global assessment. The first report of the IUCN sampled red list index for plants. Royal Botanic Gardens, Kew, UK.

Buckland, S.T., Anderson, D.R., Burnham, K.P., Laake, J.L., Borchers, D.L. and Thomas, L. (2001). Introduction to Distance Sampling: Estimating Abundance of Biological Populations. Oxford University Press, New York.

Bullock, S. H. and Solis Margallenus, J. A. (1990). Phenology of canopy trees of a tropical deciduous forest in Mexico. *Biotropica*, 22: 22-35.

Chandrashekara, U. M. (1992). Studies on Gap Phase Dynamics of a Humid Tropical Forest. Ph.D. Thesis, Jawaharlal Nehru University, New Delhi.

Cox, P. M., Betts, R. A., Jones, C. D., Spall, S. A. and Totterdell, I. J. (2000). Acceleration of global warming due to carbon-cycle feedbacks in a coupled climate model. *Nature*, 408: 750.

Champion, H. G. (1936). A preliminary survey of the forest types of India and Burma. *Indian Forest Records*, 1: 1–286.

Campbell, D.G., Stone, J.L. and Rosas, A. Jr. (1992). A comparison of the phytosociology and dynamics of three floodplain (Varzea) forest of known ages, Rio Jurua, Western Brazilian Amazon. *Botanical Journal of the Linnean Society*, 108: 231-237.

Campbell, E. J. F. and Newbery, D. Mc. (1993). Ecological relationships between lianas and trees in lowland rainforest in Sabh, East Malaysia. *Journal of Tropical Ecology*, 9: 469-490.

Carbone, C., Christie, S., Conforti, K., Coulson, T., Franklin, N., Ginsberg, J. R., Griffinths, M., Holden, J., Kawanishi, K., Kinnaird, M., Laidlaw, R., Lynam, A., Macdonald, D. W., Martyr, D., McDougal, C., Nath, L., O'Brien, T., Seidensticker, J., Smith, J.L., Sunquist, M., Tilson, R. and Wan Shahruddin, W. N. (2001). The use of photographic rates to estimate densities of tigers and other cryptic mammals. *Animal Conservation*, 4(1): 75–79.

Champion, H.G. and Seth, S.K. (1968). A revised Survey of the Forest Types of India. Govt. of India Press, Delhi.

Chandrashekara, U. M. and Ramakrishnan, P. S. (1994). Vegetation and gap dynamics of a tropical wet evergreen forest in the western ghats of Kerala, India. *Journal of Tropical Ecology*, 10: 337-354.

Chauhan, D.S., Singh, R., Mishra, S., Dadda, T., Goyal, S.P., (2006). Estimation of Tiger Population in an Intensive Study area of Pakke Tiger Reserve, Arunachal Pradesh, India. Wildlife Institute of India, Dehradun, India.

Chittibabu, C.V. and Parthasarathy, N. (2000). Attenuated tree Species diversity in human-impacted tropical evergreen forest sites at Kolli hills, eastern ghats, India. *Biodiversity and Conservation,* 9: 1493-1519.

Christelle, F. G., Doumunge, C., Mc Key, D., Tchouto, P. M. G., Sunderland, T. C. H., Balinga, M. P. B. and Snoke, B. (2011). Tree diversity and conservation value of Ngovayang's lowland forests, Cameroon. *Biodiversity and Conservation,* 20: 2627-2648.

Chuyong , G. B., Kenfack, D., Harms, K. E., Thomas, D. W., Condit, R. and Comita, L. S. (2011). Habitat specificity and diversity of tree species in an African wet tropical forest. *Plant Ecology,* 212: 1363-1374.

Cleary, D.F.R., Boyle, T.J.B, Setyawati,T., Anggraeni, C.D., Van Loon, E.E. and Menken, S.B.J. (2007). Bird species and traits associated with logged and unlogged forest in Borneo. Ecological Applications, 17:1184-1197.

Condit, R., Hubbell, S. P. and Foster, R. B. (1996). Assessing the response of plant functional types to climatic change in tropical forests. *Journal of Vegetation Science,* 7: 405-416.

Couteron, P., Pelissier, R., Nicolini, E. A. and Paget, D. (2005). Predicting tropical forest stand parametres from fourier transform of very high-resolution remotely sensed canopy images. *Journal of Applied Ecology,* 42: 1121-1128.

Curtis, J. T. and Mc Intosh, R. P. (1950). The interrelations of certain analytical and synthetic phytosociological characters. *Ecology,* 31: 434-455.

Cutler, T. L. and Swann, D. E. (1999). Using remote photography in wildlife ecology: a review. *Wildl. Soc. Bull.* , 27: 571–581.

Datta, A., Anand, M. O. and Naniwadekar, R. (2008). Empty forests: large carnivore and prey abundance in Namdapha national park, north-east India. *Biological Conservation,* 141: 1429-1435.

Datta, N. P., Khera, M. S. and Saini, T. R. (1962). A rapid colorimetric procedure for the determination of the organic carbon in soils. *Journal of Indian Society of Soil Science,* 10: 67-74.

Davidar, P., Arjunan, M., Mammen, P. C., Garrigues, J. P. and Puyravaud, J. P. (2007). Forest degradation in the western ghats biodiversity hotspot: In resource collection, livelihood concerns and sustainability. *Current Science,* 93: 1573–157.

Davidar, P., Sahoo, S., Mammen, P. C., Acharya, P., Puyravaud, J. P., Arjunan, M., Garrigues, J. P. and Roessingh, K. (2010). Assessing the extent and causes of forest degradation in India: In where do we stand? *Biological Conservation,* 143: 2937-2944.

Dixit, A.M. (1997). Ecological evaluation of dry tropical forest vegetation: An approach to environmental impact assessment. *Tropical Ecology,* 38: 87-100.

Dixon, J.A. and Sherman, P. B. (1990). Economics of Protected Areas: A New Look at Benefits and Costs. Covelo, Calif Island Press.

Duffy, J.E. (2003). Biodiversity loss, trophic skew, and ecosystem functioning. *Ecology Letters*, 6(8): 680-687.

Dutta, S.K., Nair, M. V., Mohapatra, P. P. and Mohapatra, A.K. (2009). Amphibians and Reptiles of Similipal Biosphere Reserve. Plant Resource Centre, Bhubaneswar.

Dzwonko, Z. and Gawronski, S. (2002a). Effect of litter removal on species richness and acidification of mixed oak pine woodland. *Biological Conservation*, 106: 389-398.

Dzwonko, Z. and Gawronski, S. (2002b). Influence of litter and weather on seedling recruitment in a mixed oak pine woodland. *Annals of Botany*, 90: 2454-251.

Edwards, C.A. and Heath, G.W. (1963). The role of soil animals in breakdown of leaf material. In: Doeksen, J. and Van der Drift, J. (Eds). Soil Organisms. North Holland Publishing Co., Amsterdam.

Elliott, J. A., Jones, I. D. and Thackeary, S. J. (2006). Testing the sensitivity of phytoplankton communities to changes in water temperature and nutrient load, in a temperate lake. *Hydrobiologia*, 559: 401–411.

Facelli, J.M. and Pickett, S.T.A. (1991a). Plant litter: Its dynamics and effects on plant community structure. *Botanical Review*, 57: 1-32.

Facelli, J. M. and Pickett, S. T. A. (1991 b). Plant litter: Its dynamics and effects on plant community structure. *The Botanical Review*, 57: 1-32.

FAO (1988). Food and Agriculture organization. Soil Map of the World. Revised Legend. Reprinted with corrections. *World Soil Resources Report* 60. FAO, Rome.

FAO (1990). Food and Agriculture organization. Guidelines for Soil Profile Description. Third edition (revised). Soil Resources, Management and Conservation Service, Land and Water Development Division, FAO, Rome.

FAO (2001). Food and Agriculture organization. State of The Worlds Forests., Rome.

Felfili, J., Nascimento, A. R. T., Fagg, C. W. and Meirelles, E. A. (2007). Floristic composition and community structure of a seasonally deciduous forest on limestone outcrops in Central Brazil. *Revista Brasileria de Botanica*, 30(4): 611-621.

Foster, R.B. (1996). The seasonal rhythm of fruit fall in Barro Colorado Island. 7-12 In: Leigh,E.G. Jr. (Ed.). The Ecology of a Tropical Forest. 2nd Edition. Smithsonian Inst. Press, Washington, DC.

Fox, J.E.D. (1976). Constraints on the natural regeneration of tropical moist forests. *Forest Ecology and Management*, 1: 37-65.

Frankie, G.W., Baker, H. G. and Opler, P.A. (1974). Comparative phenological studies of trees in tropical wet and dry forest in the low lands of Costa Rica. *Journal of Ecology*, 62: 881-913.

FSI (Forest Survey of India). (1999). The State of Forest Report. Forest Survey of India, Dehradun, India.

FSI (Forest Survey of India). (2006). The State of Forest Report, Forest Survey of India, Dehradun, India.

FSI (Forest Survey of India). (2011). The State of Forest Report, Forest Survey of India, Dehradun, India.

Geist, H.J. and Lambin, E. F. (2002). Proximate causes and underlying driving forces of tropical deforestation. *BioScience*, 52: 143–150.

Gholz, H.L., Wedin, D.A., Smitherman, S.M., Harmon, M.E. and Parton, W.J. (2000). Long-term dynamics of pine and hardwood litter in contrasting environments: toward a global model of decomposition. *Global Change Biology*, 6: 751-765.

Ginter, D. l., McLeod, K. W. and Sherrod, C. (1979). Water stress in longleaf pine induced by litter removal. *Forest Ecology and management*, 2: 13-20.

Givinish, T. J. (2002). Adaptive significance of evergreen vs. deciduous leaves: solving the triple paradox. *Silva Femina*, 36:703-743.

Godoy, R., Groff, S. and O'Neill, K. (1997). The Role of Education in Neotropical Deforestation: Household Evidence from Amerindians in Honduras. Harvard Institute for International Development, Harvard University. Mimeo.

González, G. and Seastedt, T.R. (2001). Soil fauna and plant litter decomposition in tropical and subalpine forests. *Ecology*, 82: 955-964.

González, G. and Zou, X. (1999). Plant litter influences on earthworm abundance and community structure in a tropical wet forest. *Biotropica*, 31(3): 486- 493.

Gosz, J.R., Likens, G.E. and Bormann, F.H. (1973). Nutrient release from decomposing leaf and branch litter in the Hubbard Brook Forest, New Hampshire. *Ecological Monographs*, 43: 173-191.

Grime, J. P. (1979). Plant Strategies and Vegetation Processes. John Wiley and Sons, Chichester.

Haines, H.H. (1921-1925). The Botany of Bihar and Odisha. Vol. I-III. London, Botanical Survey of India, Calcutta (Repn. Edn. 1961).

Haines, H.H. (1925). The Botany of Bihar and Odisha. Vol. I-III. London, Botanical Survey of India, Calcutta (Repn.Edn. 1961).

Hamann, A. (2004). Flowering and fruiting phenology of a Philippine submontane rainforest: climatic factors as proximate and ultimate causes. *Journal of Ecology*, 92: 24-31.

Hare, M. A., Lantage, D. O., Murphy, P. G. and Checo, H. (1997). Structure and tree species composition in a subtropical dry forest in the Dominican Republic: Comparison with a dry forest in Puerto Rico. *Tropical Ecology*, 38 (1): 1-17.

Heaney, A. and Proctor, J. (1990). Preliminary studies of forest structure and floristics on volcan barva, Costa Rica. *Journal of Tropical Ecology*.

Heinrich, A. and Hurka, H. (2004). Species richness and composition during sylvigenesis in a tropical dry forest in north western Costa Rica. *Tropical Ecology*, 45(1): 43-57.

Heneghan, L., Coleman, D.C., Zou, X., Crossley, D.A. and Haines, B.L. (1999). Soil micro arthropod contributions to decomposition dynamics: Tropical-temperate comparisons of a single substrate. *Ecology*, 80: 1873-1882.

Hewit, N. and Kellman, M. (2002). True seed dispersal among forest fragments: dispersal ability and biogeographical controls. *Journal of Biogeography*, 29 (3): 351 -363.

Hobbie, S.E. (1996). Temperature and plant species control over litter decomposition in Alaskan tundra. *Ecological Monographs*, 66 (4): 503– 522.

IUCN, (2013). IUCN red list of threatened species. Version 2013.2. <http:// www. iucnredlist.org>.

Jackson, M. L. (1973). Soil Chemical Analysis. Prentice hall of India Pvt. Ltd., New Delhi.

Janzen, D.H . (1970). Hervivores and the number of trees species in tropical forest. *Am. Nat.*, 104:501-528.

Jathanna, D., Karanth, K. U. and Johnsingh, A. J. T. (2003). Estimation of large herbivore densities in the tropical forests of southern India using distance sampling. *J. Zool.*, 261: 285–290.

Jennelle, C.S., Runge, M.C. and Mackenzie, D.I. (2002). The use of photographic rates to estimate densities of tigers and other cryptic mammals: a comment on misleading conclusions. *Animal Conservation*, 5: 119-120.

Jordan, C. F. (1971). A world pattern in plant energetics. *American Science*, 59: 426-433.

Justiniano, M.J. and Fredericksen, T. S. (2000). Phenology of timber tree species in a Bolivian dry forest: implications for forest management. *Journal of Tropical Forest Science*, 12(1): 174-180.

Karanth, K. K., Nichols, J. D., Hines, J. E., Karanth, U. K. and Christensen, N. L. (2009). Patterns and determinants of mammal species occurrence in India. *J. Appl. Ecol.* ,46: 1189-1200.

Karanth, K. K., Nichols, J. D., Karanth, K. U., Hines, J. E. and Christensen, N. L. (2010). The shrinking ark: Patterns of large mammal extinctions in India. *Proc. R. Soc. London*, 277: 1971-1979.

Karanth, K. U. and Nichols, J. D. (1998). Estimation of tiger densities in India using photographic captures and recaptures. *Ecology*, 79: 2852–2862.

Karanth, K.K., Curran, L. M. and Reuning-Scherer, J. D. (2006). Village size and forest disturbance in Bhadra wildlife sanctuary, western ghats, India. *Biological Conservation*, 128: 147-157.

Karanth, K.U. (1995). Estimating Panthera tigris populations from camera-trap data using capture–recapture models. *Biol. Conserv.*, 71:333-338.

Kays, R.W., Slauson, K.M. (2008). Remote cameras. In: Long, R.A., MacKay, P., Zielinski, W.J., Ray, J. (Eds.), Noninvasive Survey Methods for Carnivores. Island Press, Washington.

Kershaw, K.R. (1973). Quantitative and Dynamic Plant Ecology. Edward Arnold Ltd., London.

Khumbongmayum, A.D., Khan, M. L. and Tripathi, R. S. (2006). Biodiversity conservation in sacred groves of Manipur, northeast India: population structure and regeneration status of woody species. *Biodiversity and Conservation*,15: 2439-2456.

Kigmo, B. N., Savill, P. S. and Woodell, S. R. (1990). Forest composition and its regeneration dynamics: a case study of semi-deciduous tropical forest in Kenya. *African Journal of Ecology*, 28 (3): 174-187.

Kikim, A. and Yadava, P. S. (2001). Phenology of tree species in subtropical forests of Manipur in northeast India. *Tropical Ecology*, 42(2): 269-276.

Kilian, W. (1998). Forest site degradation-temporary deviation from the natural site potential. *Ecological Engineering* ,10: 5-18.

Knight, D.H. (1975). A phytosociological analysis of species rich tropical forest on Barro Colorado Island, Panama. *Ecological Monographs*, 45: 259-289.

Knoepp Jennifer, D., C., Reynolds Barbara, D. A. and Swank Wayne, T. (2005). Long-term changes in forest floor processes in southern Appalachian forests. *Forest Ecology and Management*, 220: 300-312.

Kononova, M.M. (1975). Humus of virgin and cultivated soils. In: Gieseking, J.E. (Ed.). Soil Components, Vol. I. Springer-Verlag, New York.

Kothari, A., Pande, P., Singh, S. and Variava, D. S. (1989). Management of National Parks and Sanctuaries in India: A Status Report. IIPA, New Delhi.

Kramer. R., Healy, R. and Mendelsohn, R. (1992). Forest valuation In: Sharma, N. P. (ed.) Managing the Wandor's Forests: Looking for Balance between Conservation and Development. Kendall/Hunt Publishing Company, Dubuque.

Krishnankutty, N., Chandrasekaran, S. and Jeyakumar, G. (2006). Evaluation of disturbance in a tropical dry deciduous forest of Alagar Hill (Eastern Ghats), south India. *Tropical Ecology*, 47 (1) : 47-55.

Lacerda, A.C.R., Tomas, W.M. and Marinho, J. (2009). Domestic dogs as an edge effect in the Brasilia national park, Brazil: interactions with native mammals. *Anim. Conserv.*, 12: 477-487.

Lang, G.E. (1974). Litter dynamics in a mixed oak forest on the New Jersey Piedmont. *Bulletin of Torrey Botanical Club*, 101: 277-286.

Lawton, J.R.S. and Akpan, E.E.J. (1968). Periodicity in *Plumeria*. Nature, 218:384–386.

Leigh, E.G., Davidar, P., Dick, C.W., Puyravaud, J.P., Terborgh, J., Steege, H. and Wright, S.J. (2004). Why do some tropical forests have so many species of trees? *Biotropica*, 36: 447-473.

Leigh, E.G., Wright, S.J., Herre, E.A., and Putz, F.E. (1993). The decline of tree diversity on newly isolated tropical islands: a test of a null hypothesis and some implications. *Evolutionary Ecology*, 7: 76-102.

Leith, H. (1974). Phenology and Seasonality Modelling. Springer, Berlin.

Levin, D. A. and Anderson, W. W. (1970). Competition for pollinators between simultaneously flowering species. *Am. Nat.*, 104: 455–467.

Liberman, D. (1982). Seasonality and phenology in a dry tropical forest in Ghana. *Journal of Ecology*, 70: 791–806.

Lewis, O. T. (2009). Biodiversity change and ecosystem function in tropical forests. *Basic and Applied Ecology*, 10 (2): 97-102.

Liu, X., Wu, P., Songer, M., Cai, Q., He, X., Zhu, Y. and Shao, X. (2013). Monitoring wildlife abundance and diversity with infra-red camera traps in Guanyinshan nature reserve of Shaanxi Province, China. *Ecological Indicators*, 33: 121-128.

Lloyd, J. and Taylor, J. A. (1994). On the temperature dependence of soil respiration. *Functional Ecology*, 8: 315–323.

Loomis, R.M. (1975). Animal changes in forest floor weights under a southeast Missouri oak stand. USFS Research Note, NC- 184.

Larson, B. C. and Zaman, M. N. (1986). Thinning guidelines for teak (*Tectona grandis*). *Malay Forester*, 48(4): 288-297.

Lawson, G. W. (1985). Plant Life in West Africa. Ghana Universities Press, Accra.

Leisher, C., Sanjayan, M., Blockhus, J., Kontoleon, A. and Larsen, S. N. (2010). Does conserving biodiversity work to reduce poverty? A state of knowledge review. The Nature Conservancy, Washington, DC.

Loveridge, A.J., Wang, S.W., Frank, L.G., Seidensticker, J. (2010). People and wild felids: conservation of cats and management of conflicts. In: Macdonald, D.W., Loveridge, A.J. (Eds.), Biology and Conservation of Wild Felids. Oxford University Press, Oxford, UK. pp. 161-195.

Ludwig, J. A. and Reynolds, J. F. (1988). Statistical Ecology: A Primer on Methods and Computing. Wiley-Inter science, New York.

Mackenzie, D.I. (2005). What are the issues with 'presence/absence' data for wildlife managers? *Journal of Wildlife Management*, 69: 849–860.

Madge, D.S. (1965). Leaf fall and litter disappearance in a tropical forest. *Pedobiologia*, 5: 273-288.

Madhusudan, M.D. (2004). Recovery of wild large herbivores following livestock decline in a tropical Indian wildlife reserve. *J. Appl. Ecol.* ,41: 858–869.

Madhusudan, M.D. (2005). The global village: linkages between international coffee markets and grazing by livestock in a south Indian wildlife reserve. *Conservation Biology*, 19: 411–420.

Madhusudan, M.D. and Mishra, C. (2003). Why big, fierce animals are threatened: conserving large mammals in densely populated landscapes. In: Saberwal, V. and Rangarajan, M. (Eds.) Battles over nature: science and the politics of conservation: New Delhi.

Magurran, A. E. (1988). Ecological Diversity and its Measurement. New Jersey, Princeton University Press.

Magurran, A. E. and Henderson, P. A. (2003). Explaining the excess of rare species in natural species abundance distributions. *Nature,* 422: 714-716.

Malaisse, W. J., Sener, A. and Mahy, M. (1974). Eur. J. Biochem. 47: 365-370.

Mammen, P.C. (2007). Forest People Interface: A Rapid Assessment of Socio-Economic Profile, Resource Use Patterns of Sigur Plateau, Nilgiri Biosphere Reserve, Tamil Nadu. M.Sc. Dissertation. Pondicherry University, India.

Margalef, R. (1958). Information theory in ecology. *General Systematics,* 3: 36-71.

Marshall, T. J., Holmes, J. W. and Rose, C. W. (1996). Soil Physics. Cambridge University press, Cambridge.

Martinez-Yrizar, A. and Sarukhan , J. (1990). Litter fall patterns in a tropical deciduous forest in Mexico over a five year period. *Journal of Tropical Ecology,* 6: 433-444.

Mathai, M. (1999). Habitat Occupancy Across Anthropogenic Disturbances by Sympatric Ungulate Species in Panna Tiger Reserve. M. Sc. Thesis, Saurashtra University, Rajkot, India.

Mayers, N. (1992). The Primary Source: Tropical Forests and Our Future. Norton and Co., New York.

Mc Claugherty, C.A. and Berg, B. (1987). Cellulose, lignin and nitrogen concentrations as regulating factors in late stages of forest litter decomposition. *Pedobiologia,* 30: 101-112.

Mishra, R. K., Bal, S. and Mohanty, R. C. (2002). Phytosociological and plant diversity studies of Similipal biosphere reserve. Proceedings of the National Seminar on Conservation of Eastern Ghats, EPTRI, Tirupati, pp., 16-26.

Mishra, R. K., Mohapatra, P. K., Upadhyay, V. P., Bal, S. and Mohanty, R. C. (2001). Community structure and composition of tree layer in different altitudes of southern core areas of Similipal biosphere reserve. Proceedings of the Xth National Symposium on Envirnment, Barc, Mumbai, Pp., 223-227.

Mishra, R. K., Mohapatra, P. K., Upadhyay, V. P., Bal, S. and Mohanty, R. C. (2007). Turnover rate of forest floor litter and biodiversity aspects of Similipal biosphere reserve. In: Trivedi, P. C. (Ed.). Biodiversity for Sustainable Development. Aavishkar Publishers and Distributors, Jaipur.

Mishra, R. K., Upadhyay, V. P. and Mohanty, R. C. (2003). Vegetation diversity of Similipal biosphere reserve. *E-planet,* I: 4-9.

Mishra, R. K., Upadhyay, V. P., Pattanaik, S., Nayak, P. K. and Mohanty, R. C. (2012). Composition and stand structure of tropical moist deciduous forest of Similipal biosphere reserve, Orissa, India. In: Blanco, J. A. and lo, Y. H. (Eds.). *Forest Ecosystems More Than Just Trees,* InTech Publisher, Croatia.

Mishra, B. P., Tripathi, O. P. and Laloo, R. C. (2005). Community characteristics of a climax subtropical humid forest of Meghalaya and population structure of ten important tree species. *Tropical Ecology,* 46: 241-251.

Mishra, R. K., Upadhyay, V. P., Bal , S., Mohapatra, P. K. and Mohanty, R. C. (2006). Phenology of species of moist deciduous forest sites of Similipal biosphere reserve. *Lyonia*, 11(1): 5-17.

Mishra, R. K., Upadhyay, V. P. and Mohanty, R. C. (2008). Vegetation ecology of the Similipal biosphere reserve, Odisha, India. *Applied Ecology and Environmental Research*, 6(2):89-99.

Misra, R. (1968). Ecological Workbook. Oxford and IBH Publishing Company, Calcutta.

Misra, R. and Puri, G. S. (1954). Indian Manual of Plant Ecology. English Book Depot, Dehradun

Mo, J., Brown, S. , Peng, S. and Kong, G. (2003). Nitrogen availability in disturbed, rehabilitated and mature forests of tropical China. *Forest Ecology and Management*, 175: 573-583.

Mohanty, R. C. (2001). Long-term Biodiversity Assessment of Similipal Biosphere Reserve. Report Submitted to Ministry of Environment and Forests, Govt. of India. Department ofBotany, Utkal University, Bhubaneswar.

Mohanty, R. C., Mishra, R. K., Bal, S. and Upadhyay, V. P. (2005). Plant diversity assessment of Shorea robusta dominated forest sites of Similipal biosphere reserve. *Journal of Indian Botanical Society*, 84(1-4): 21-29.

Mohapatra, P.P., Palei, H.S. and Hussain, S.A. (2014). Occurrence of Asian small-clawed otter Aonyx cinereous (Illeger, 1815) in Eastern India, *Current Science*, 107:367-370.

Monasterio, M. and Sarmiento, G. (1976). Phenological strategies of plant species in the tropical savanna and semi-deciduous forest of the Venezuelan Llanos. *Journal of Biogeography*, 3: 325-356.

MacKenzie, D.I., Nichols, J.D., Lachman, G.B., Droege, S., Royle, J.A., Langtimm, C.A. (2002). Estimating site occupancy rates when detection probabilities are less than one. *Ecology*, 83: 2248–2255.

Morruzzi, T.L., Fuller, T.K., De Graaf, R.M., Brooks, R.T. and Li, W. (2002). Assessing remotely triggered cameras from surveying carnivore distribution. *Wildlife Society Bulletin*, 30: 380–386.

Mourão, G., Coutinho, M., Mauro, R., Campos, Z., Toma´s, W. and Magnusson, W. (2000). Aerial surveys of caiman, marsh deer and pampas deer in the Pantanal wetlands of Brazil. *Biological Conservation*, 92: 175–183.

Mueller-Dombois, D. and Ellenberg, H. (1974). Aims and Methods of Vegetation Ecology. John Wiley and Sons, New York.

Murali, K.S. and Sukumar, R. (1994). Reproductive phenology of a tropical dry forest in Mudumalai, Southern India. *Journal of Ecology*, 82: 759-767.

Murthy, M. S. R., Sudhakar, S., Jha, C. S., Reddy, S., Pujar, G. S. and Roy, P. S. (2007). Vegetation, land cover and phytodiversity characterisation and landscape level using satellite remote sensing and geographic information system ine eastern ghats, India. *EPTRI-ENVIS Newsletter*,13 (1): 1-12.

Nadkarni, N. M., Matelson, T.S. and Haber, W.A. (1995). Structural characteristics and floristic composition of neotropical cloud forest, Monte Verde, Costa Rica. *Journal of Tropical Ecology*, 11: 481-495.

Nascimento, A. R. T., Felfili, J. M. and Meirelles, E. M. L. (2004). Floristica e estrutura de um remanscent de floresta estucional decidual de Encosta no Municipio De Monte Alegre, GO, Brazil. *Acta Botanica Brasilica*, 18: 659-699.

Nayak, A. K. and Naik, S. (2014).Birds of Similipal Biosphere Reserve, Similipal Tiger Reserve, Baripada, Odisha.

Newbery, D. Mc C., Kennedy, D. N., Petol, G. H., Madani, L. and Ridsdale, C. E. (1999). Primary forest dynamics in lowland Dipterocarp Forest at Danum Valley, Sabh, Malaysia and the role of understorey. Philosophical Transactions of Royal Society of London, Series B, 354:1763-1782.

Newbery, D. McC. (1991). Floristic variation within Kerangas (Heath) Forest: revaluation of data from Sarawak and Brunei. *Vegetatio*, 96: 43-86.

Nichols J.D. (1992). Capture-recapture models. *Bioscience*, 42:94–102.

Nirmal Kumar, J. I., Kumar, R. N., Bhoi, R. K. and Sajish, P. R. (2010). Tree species diversity and soil nutrient status in three sites of tropical dry deciduous forest of western India. *Tropical Ecology*, 51 (2): 273-279.

Njoku, E. (1964). Seasonal periodicity in the growth and development of some forest trees in Nigeria: observations on seedlings. *Journal of Ecology*, 52:19–26.

O'Brien, T.G., Kinnaird, M.F., and Wibisono, H.T. (2003). Crouching tigers, hidden prey: Sumatran tiger and prey populations in a tropical forest landscape. *Animal Conservation*, 6:131–139.

O'Connell, A.F., Nichols, J.D. and Karanth, K.U. (2011). Camera Traps in Animal Ecology: Methods and Analyses. Springer.

Odum, E.P. (1971). Fundamentals of Ecology. W.B. Saunders Co., Philadelphia.

Ogee,J. and Brunet, Y. (2002). A forest floor model for heat and moisture including a litter layer. *Journal of Hydrology*, 255: 212-233.

Olsen, S. R., Cole, C. V., Watanabe, F. S. and Dean, L. A. (1954). Estimation of available phosphorus in soils by extraction with sodium bicarbonate. *Circ. U.S. Dept. Agric.*

Olson, J. S. (1963). Energy storage and the balance of producers and decomposers in ecological systems. *Ecology*, 44: 322-331.

Opler, P.A., Frankie, G. W. and Baker, H. G. (1980). Comparative phonological studies of shrubs and treelets in wet and dry forests in the lowlands of Costa Rica. *Journal of Ecology*, 68: 167-186.

Oyun, M. B., Bada, S. O. and Anjah, G. M. (2009). Comparative analysis of the floral composition at the edge and interior of Agulii forest reserve, Cameroon. *Journal of Biological Science*, 9(5):431-437.

Padalia, H., Chauhan, N., Porwal, M.C. and Roy, P.S. (2004). Phytosociological observations on tree species diversity of Andaman Islands, India. *Current Science*, 87: 799-806.

Pandey, U. and Singh, J. S. (1981). A quantitative study of the forest floor, litter fall and nutrient return in an oak-conifer forest in Himalaya. I. Composition and dynamics of forest floor. *Acta Oecologia*, 2: 49-61.

Parida, R.C. (1997). Sustainable exploitation of natural resources of Similipal. In: Tripathy, P.C. and Patro, S.N. (Eds.) Similipal: A Natural Habitat of Unique Biodiversity. Odisha Environmental Society, Bhubaneswar.

Parker, J. S. (1984). UNESCO documents and publications in the field of information: a summary guide. *IFLA Journal*, 10(3): 251-272.

Parthasarathy, N. (1999). Tree diversity and distribution in undisturbed and human-impacted sites of tropical wet evergreen forest in southern western ghats, India. *Biodiversity and Conservation*, 8: 1365-1381.

Parthasarathy, N. and Karthikeyan, R.(1997). Biodiversity and population density of woody species in a tropical evergreen forest in Courtallum reserve forest, western ghats, India. *Tropical Ecology*, 38: 297-306.

Pascal, L. P., Pellissier, R. (1996). Structure and floristic composition of a tropical evergreen forest in south-west India. *Journal of Tropical Ecology*, 12: 191- 210.

Pascal, J. P. (1984). Forest humids sempervirentes des Ghats occidentaux de inde. Institute Francals d' Pondicherry.

Patnaik, S. K. (1997). Conservation and management of wildlife in Similipal. In: Tripathy, P.C. and Patro, S.N. (Eds.). Similipal: A Natural Habitat of Unique Biodiversity. Orissa Environmental Society, Bhubaneswar.

Paustian, K., Six, J., Elliott, E. T. and Hunt, H. W. (2000). Management options for reducing CO_2 emissions from agricultural soils. *Biogeochemistry*, 48: 147–163.

Pelissier, R., Pascal, J.P., Houllire, F. and Laborde, H. (1998). Impacts of selective logging on the low elevation dense moist evergreen forest in the western ghats (South India). *Forest Ecology and Management* ,105:107-119.

Peres, C.A. (2000). Effects of subsistence hunting on vertebrate community structure in Amazonian forests. *Conserv. Biol.*, 14: 240–253.

Peres, C.A., Dolman, P.M. (2000). Density compensation in neo tropical primate communities: evidence from 56 hunted and non-hunted Amazonian forests of varying productivity. *Oecologia*, 122: 175–189.

Persha, L., Agrawal, A., and Chhatre, A. (2011). Social and ecological synergy: local rule making, forest livelihoods and biodiversity conservation. *Science*, 331: 1606-1608.

Pielou, E.C. (1975). Ecological Diversity. John Wiley and Sons, New York.

Plumptre, A.J., Harris, S. (1995). Estimating the biomass of large mammalian herbivores in a tropical montane forest: a method of faecal counting that avoids assuming a 'steady state' system. *Journal of Applied Ecology*, 32: 111–120.

Pollock, K.H., Nichols, J.D., Simons, T.R., Farnsworth, G.L., Bailey, L.L. and Sauer, J.R. (2002). Large-scale wildlife monitoring studies: statistical methods for design and analysis. *Environmetrics*, 13: 105–119.

Polunin, N. V. C. (1984). The decomposition of emergent macrophytes in fresh water. *Advances in Ecological Research*, 14: 115-166.

Ponge, J. F., Arpin, P. and Vannier, G. (1993). Collembolan response to experimental perturbations of litter supply in a temperate forest ecosystem. *European Journal of Soil Science*, 29: 141-153.

Poulsen, A.D., Nielsen, I.V. and Balslev, H. (1996). A quantitative inventory of trees in one hectare of mixed Dipterocarp forest in Temburong, Brunei Durussalam, In: Edwards, D.S., Booth, W.E. and Choy , S.C. (Eds.) Tropical Rain Forest Research-Current Issues, Kluwer Academic Publishers, Dordrecht, The Netherlands.

Pragasan, L. A. and Parthasarathy, N. (2005). Litter production in tropical dry evergreen forests of south India in relation to season, plant life forms and physiognomic groups. *Current Science*, 88(8): 1255-1263.

Puri, G. S. (1995). Biodiversity and development of natural resources for the 21st Century. *Tropical Ecology*, 56 (2): 253-255.

Puyravaud, J. P. and Garrigues, J. P. (2002). L'agriculteur ou la forêt? Systèmes agraires, prélèvements et conséquences écologiques sous la crête des Ghâts (district de Shimoga). In L'homme et la forêt en Inde du Sud. In: Pouchepadess, J. and Puyravaud, J. (Eds.) Modes de gestion et symbolisme de la forêt dansles Ghâts occidentaux, Institut Françaisde Pondichéry–Karthala.

Rai, B. and Srivastava, A. K. (1982). Litter production in a tropical dry mixed forest stand. *Acta Oecol.*, 3: 169-176.

Rai, S. N. and Proctor, J. (1986). Ecological studies on four rainforests in Karnataka, India: II. Litterfall. *Journal of Ecology*, 74: 455-463.

Rajendraprasad, P., Nageshwara Rao, K., Prasad, N .V . B. S.S., Raman, A,V., Bhanu Kumar, O.S.R.U.,Swamy,Y,V., Das, S,N., Chaudhury, G,R., Kumar, A.R.S., Sudhakar, S.V., Rama Rao, P., Jagannadha Rao, S., Sampath Kumar, M., Deva Varma, D., Bhanumurthy, P., Ganesh, T., Somu Naidu, Y. Sanjeevaiah, G., Mary Padma, V. and Lokesh Kumar Reddy, N. (2006). Effect of 26 December 2004 tsunami along the Andhra coast- An integrated study. In: Rajmanickam et al., 2006 (Eds.). Earth System Science, Department of Science & Technology, Government of India, New Delhi.

Rajendraprasad, M. (1995). The Floristic, Structural and Analysis of Sacred Groves of Kerala. Ph. D. Thesis, Kerala University, Thiruvanthapuram.

Rajendraprasad, M., Krishnan, P. N. and Pushpangadan, P. (2000). Vegetational characterization and litter dynamics of the sacred grooves of Kerala, southwest India. *Journal of Tropical Forest Science*, 12(2): 320-325.

Ralhan, P.K., Khanna, R. K., Singh, S. P. and Singh, J.S. (1985a). Phenological characteristics of the tree layer of Kumaun Himalayan forests. *Vegetatio*, 60: 91-101.

Ralhan, P.K., Khanna, R. K., Singh, S. P. and Singh, J. S. (1985b). Certain phonological characters of the shrub layer of Kumaun Himalayan forests. *Vegetatio*, 63: 113-120.

Rath, B. and Sutar, P.C. (2004). Human Issues in Protected Areas: A Case Study in Similipal Tiger Reserve.Vasundhara.

Raubenheimer, D. and Simpson, S. J. (1999). Integrating nutrition: a geometrical approach. *Entomologia Experimentalise Applicata* , 91: 67-82.

Raunkiaer, C. (1934). The Life Form of Plants and Statistical Plant Geography. Claredon Press, Oxford.

Raut, D.K. and Behera, G. (1997). Application of remote sensing for management of natural resources in Similipal biosphere reserve. In : Tripathy, P.C. and Patro, S.N. (Eds.) Similipal: A Natural Habitat of Unique Biodiversity. Odisha Environmental Society, Bhubaneswar.

Reddy, C. S., Pattanaik, C., Mohapatra, A. and Biswal, A. K. (2007).Phytosociological observations on tree diversity of tropical forest of Similipal biosphere reserve, Odisha, India. Taiwania, 52(4): 352-359.

Reich, P. B. and Borchert, R. (1984). Water stress and tree phenology in a tropical dry forest in the low lands of Costa Rica. *Journal of Ecology*, 72: 61-74.

Roessingh, K. (2006). Mapping the Elephant Corridors of Sigur Plateau, Nilgiri Biosphere Reserve, Tamilnadu. M.Sc. Dissertation, Pondicherry University, India.

Rout, S. D., Panda, S. K., Mishra, N. and Panda, T. (2010). Role of tribals in collection of commercial non-timber forest products in Mayurbhanj district, Odisha. *Journal of Studies of Tribes and Tribals*, 8(1): 21-25.

Rout, S. D., Panda, T. and Mishra, N. (2009). Ethnomedicinal studies on some pteridophytes of Similipal biosphere reserve, Odisha, India. *International Journal of Medicine and Medical Sciences*, 1(5) :192-197.

Rovero, F. and Marshall, A. R. (2008). Camera trapping photographic rate as an index of density in forest ungulates. *Journal of Applied Ecology*, 46: 1011–1017.

Roy, P. S., Dutt, C. B. S. and Joshi, P. K. (2002). Tropical forest resource assessment and monitoring. *Tropical Ecology*, 43(1): 21-37.

Sagar, S.R. and Singh, L.A.K. (1990). Technique to distinguish tracks of leopard and tiger cubs. *Indian Forester*, 116(3): 214-219.

Sahoo, S. and Davidar, P. (2013). Effect of harvesting pressure on plant diversity and vegetation structure of sal forests of Similipal tiger reserve, Odisha. *Tropical Ecology*, 54: 97-107.

Sapkota, I.P., Tigabu, M. and Oden, P. C. (2009). Species diversity and regeneration of old-growth seasonally dry *Shorea robusta* forests following gap formation. *Journal of Forestry Research*, 20 :7-14.

Santapau, H. (1962). The Flora of Saurashtra. Part- I. Saurashtra Research Society, Rajkot

Saranya, K. R. L., Reddy, C. S., Prasad Rao, P.V.V. and Jha, C. S. (2014). Decadal time scale monitoring of forest fires in Similipal biosphere reserve, India using remote sensing and GIS. Environmental Monitoring and Assessment, 186: 3283-3296.

Saxena, H.O. and Brahmam, M. (1989). The Flora of Similipahar (Similipal), Odisha. Regional Research Laboratory, Bhubaneswar.

Saxena, H.O. and Brahmam, M. (1994-1996). The Flora of Orissa. Vol. I-IV. Regional Research laboratory (CSIR), Bhubaneswar and Orissa Forest Development Corporation Ltd., Bhubaneswar.

Sayer, E. J. (2006). Using experimental manipulation to assess the roles of leaf litter in the functioning of forest ecosystems. *Biological Review*, 81: 1-31.

Schaller, G. (1977). Mountain Monarchs: Wild Sheep and Goats of the Himalaya. University of Chicago Press, Chicago, USA.

Schnitzer, S. A. and Bongers, F. (2002). The ecology of lianas and their role in forests. *Trends in Ecology and Evolution*, 17(5): 223-230.

Sethi, P. and Howe, H.F. (2009). Recruitment of hornbill-dispersed trees in hunted and logged forests of the Indian Eastern Himalaya. *Conservation Biology*, 23:1-9.

Shaheen, H., Qureshi, R. A. and Shinwari, Z. K. (2011). Structural diversity, vegetation dynamics and anthropogenic impact on lesser Himalayan subtropical forests of Bagh district, Kashmir. *Pakistan Journal of Botany*, 43 (4) :1861-1866.

Shanker, K. (2000). Small mammal trapping in tropical montane forests of the Upper Nilgiris, southern India: an evaluation of capture-recapture models in estimating population size. *Journal of Biosciences*, 25: 99-111.

Shanks, R.E. and Olson, J.E. (1961). First year breakdown of leaf litter in Southern Appalachian forest. *Science*, 134: 370-376.

Shannon, C.E. and Wiener, W. (1963). The Mathematical Theory of Communication. University Press, Illinois, USA.

Sharma, S., Jhala, Y. and Sawarkar, V.B. (2005). Identifying individual tigers from their pugmarks. *Journal of Zoology*, 267: 9–18.

Sheridan, C.D. and Olson. D. H. (2003). Amphibian assemblages in zero-order basins in the Oregon coast range, *Canadian Journal of Forest Research*, 33:1452-1477.

Shukla, R.P. and Ramakrishnan, P. S. (1982). Phenology of trees in a subtropical humid forest in north eastern India. *Vegetatio*, 49: 103-109.

Silori, C. S. and Mishra, B. K. (2001). Assessment of livestock grazing pressure in and around the elephant corridors in Mudumalai wildlife sanctuary, South India. *Biodiversity and Conservation*, 10: 2181–2195.

Silva-Rodriguez, E.A., Ortega-Solis, G.R. and Jimenez, J.E. (2010). Conservation and ecological implications of the use of space by chilla foxes and free-ranging dogs in a human-dominated landscape in southern Chile. *Aust. Ecol.*, 35: 765-777.

Silva-Rodríguez, E.A. and Sieving, K.E. (2012). Domestic dogs shape the landscape-scale distribution of a threatened forest ungulate. *Biological Conservation*, 150: 103-110.

Simpson, E. H. (1949). Measurement of diversity. *Nature*, 163: 688.

Singh, G.S. and Gibson, L. (2011). A conservation success story in the otherwise dire maga fauna extinction crisis: the Asiatic lion (*Panthera leo persica*) of Gir forest. *Biological Conservation*, 144: 1753-1757.

Singh, J.S. (2002). The Biodiversity crisis: a multifaceted review. *Current Science*, 82: 638-647.

Singh, J.S. and Singh, V.K. (1992). Phenology of seasonally dry tropical forest. *Current Science*, 63(11): 684-689.

Sivaraj, N. and Krishnamurthy, K. V. (2002). Phenology and reproductive ecology of tree taxa of Shervaroy Hills, eastern ghats, south India. Proceedings of National Seminar on Conservation of Eastern Ghats, Tirupati, March, 24-26.

Spies, T.A. and Turner, M. G. (1999). Dynamic forest mosaics. In: Hunter, M. L.(Ed.) Maintaining Biodiversity in Forest Ecosystems. Cambridge University Press, Cambridge, UK.

Srivastava, S. S. and Singh, L. A. K. (1997). Monitoring of precipitation and temperature of Similipal tiger reserve, In: Tripathy, P. C. and Patro, S. N., Similipal (Eds.): A National Habitat of Unique Biodiversity, 30-40, Odisha Environmental Society, Bhubaneswar.

Stork, N.E. (1991). The composition of the arthropod fauna of Bornean lowland rain forest trees. *Journal of Tropical Ecology*, 7 (2): 160-180.

Sunderpandian, S. M. and Swamy, P. S. (1999). Litter production and leaf litter decomposition of selected tree species in tropical forests at Kodayar in the Western Ghats, India. *Forest Ecology and Management*, 123: 231-244.

Sunderpandian, S.M. and Swamy, P.S. (2000). Forest ecosystem structure and composition along an altitudinal gradient in the western ghats, south India. *Journal of Tropical Forest Science*, 12 (1): 104-123.

Sundriyal, R.C. (1990). Phenology of some temperate woody species of the Garhwal Himalaya. *International Journal of Ecology and Environmental Sciences*, 6: 107-117.

Swamy, H.R. (1989). Study of Organic Productivity, Nutrient Cycling and Small Watershed Hydrology in Natural Forests and in Monoculture Plantations in Chikmagalore District, Karnataka. Report Submitted to Ministry of Environment and Forests, Govt. of India. Sri J.C.B.M. College, Sringeri, Karnataka.

Swift, M.J. and Anderson, M. (1983). Decomposition. In: Leith, H. and Werger, M.S.A. (Eds.). Ecosystems of the world: Tropical Rainforest Ecosystems. Elsevier, Amsterdam.

Swift, M.J., Heal, O. W. and Anderson, J.M. (1979). Decomposition in Terrestrial Ecosystem. Blackwell Scientific Publication, London, UK.

Tilman, D. (2000). Causes, consequences and ethics of biodiversity. *Nature*, 405: 208-211.

Tobler, M. W., Carrillo-Percastegui, S. E., Leite Pitman, R., Mares, R. and Powell, G. (2008). An evaluation of camera traps for inventorying large- and medium-sized terrestrial rainforest mammals. *Animal Conservation*, 11: 169-178.

Toky, O.P. and Ramakrishnan, P.S. (1984). Litter decomposition related to secondary succession and species type under slash and burn agriculture (Jhum) in north eastern India. *Proceedings of the Indian National Science Academy*, 50: 57-65.

Toledo-Aceves, T. and Swaine, M. D. (2008). Effect of lianas on tree regeneration in gaps and forest understorey in a tropical forest in Ghana. *Journal of Vegetation Science*, 19 (5) :717-728.

Townsend, A. R., Asner, G. P. and Cleveland, C. C. (2008). The biogeochemical heterogeneity of tropical forests. *Trends in Ecology and Evolution*, 23 (8): 424-431.

Treves, A. P., Mwina, A. J. and Isoke, S. (2010). Camera trapping forest-woodland wildlife of western Uganda reveals how gregariousness biases estimates of relative abundance and distribution. *Biological Conservation*, 143: 521-528.

Tripathi, K. P. and Singh, B. (2009).Species diversity and vegetation structure across various strata in natural and plantation forest in Katerniaghat wildlife sanctuary, north India. *Tropical Ecology*, 50 (1): 191-200.

Tripathi, N. and Singh, R. (2012). Impact of savannization on nitrogen mineralization in an Indian tropical Forest. *Forest Research*,1(3):1-10.

Trolle, M. and Ke´ry, M. (2005). Camera-trap study of ocelot and other secretive mammals in the northern Pantanal. *Mammalia*, 69: 405-412.

Tsingalia, M. T. (1990). Habitat disturbance, sverity and patterns of abundance in Kakamega forest, western Kenya. *African Journal of Ecology*, 28: 213- 226.

UNESCO (2011). Biosphere Reserves World Network - Ecological Sciences for Sustainable Development.

UNESCO (1971). Science, Policy and the European States. Science, Policy Studies and Documents No. 25, Paris.

Upadhyay, S., Sahu, S.K., Panda, G. K. and Upadhyay, V. P. (2012). Human ecology of a village in Similipal biosphere reserve, Odisha, India. *Plant Science Research*, 34 (1 and 2): 54-59.

Upadhyay, V. P. and Singh, J S. (1989). Patterns of nutrient immobilization and release in decomposing forest litter in Central Himalaya, India. *Journal of Ecology*, 77: 127-146.

Upadhyay, V. P., Singh, J S. and Meentemeyer, A.T. (1989). Dynamics and weight loss of leaf litter in central Himalayan forest: abiotic versus litter quality influences. *Journal of Ecology*, 77: 127-146..

Valencia, R., Balslev, H. and Mino, P. (1994). High tree alpha diversity in Amazonian ecuador. *Biodiversity and Conservation*, 3 : 21-28.

Vanak, A.T., Gompper, M.E. (2010). Interference competition at the landscape level: the effect of free-ranging dogs on a native mesocarnivore. *J. Appl. Ecol.*, 47: 1225–1232.

Vanschaik, C. P. (1986). Phenological changes in a Sumatran rain forest. *Journal of Tropical Ecology*, 2: 327-347.

Vasanthraj, B. K., Shivaprasad, P. V. and Chandrashekar, K. R. (2005). Studies on the structure of Jadkal forest, Udupi district, India. *Journal of Tropical Forest Science*, 17: 13-32.

Viana, V.M. and Tabanez, A.A.J. (1996). Biology and conservation of forest fragments in the Brazilian Atlantic moist forest. In: Schelas, J. and Greenberg R. (Eds.), Forest Patches in Tropical Landscapes. Island Press, Washington.

Visalakshi, N. (1995). Vegetation analysis of two tropical dry evergreen forests in southern India. *Tropical Ecology*, 36: 117-127.

Vogt, K. A., Grier, C. C. and Vogt, D. J. (1986). Production, turnover and nutrient dynamics of above ground and below-ground detritus of world forests. *Advances in Ecological Research*, 15:303-377.

Vordzogbe, V. V., Attuquayefio, D. K. and Gbogbo, F. (2005). The flora and mammals of moist semi-deciduous forest zone in the Sefwi Wiawso district of the western region, Ghana. *African Journal of Ecology*, 8: 49-64.

Wall, D.H. and Virginia, R. A. (2000). The world beneath our feet: soil biodiversity and ecosystem functioning. In: Raven, P. R. and Williams, T. (Eds.). Nature and Human Society: The Quest for A Sustainable World. National Academy of Sciences and National Research Council, Washington, DC.

WCMC (1992). World Conservation Monitoring Centre. Global Biodiversity: Status of Earth's Living Resources. Chapman and Hall, London, UK.

Weary, G.C. and Merriam, H.G. (1978). Litter decomposition in a red maple woodlot under natural conditions and under insecticide treatment. *Ecology*, 59: 180-184.

Weber, R. (1985). Basic Content Analysis. Beverly Hills, CA: Sago

Whitford, W.G., Freckman, D.W., Santos, P.F., Elkins, N.Z. and Parker, L.W. (1982). The role of nematodes in decomposition in desert ecosystems. In: Freckman, D. W. (Ed.). Nematodes in soil ecosystems. University of Texas Press, Austin.

Wiant, H. V. (1967). Influence of temperature on the rate of soil respiration. *Journal of Forestry*, 65: 489-490.

William, S.T. and Gray, T.R.G. (1974). Decomposition of litter on the soil surface. In: Dickinson, C.H. and Pugh, G.J.F. (Eds.). Biology of Plant Litter Decomposition. Academic Press, London.

Weidelt, H. J. (1988). On the diversity of tree species in tropical rain forest eosystems. *Plant Research and Development*, 28: 110-125.

Whisson, D. A., Engeman, R. M. and Collins, K. (2005). Developing relative abundance techniques (RATs) for monitoring rodent populations. *Wildl. Res.*, 32:239–244.

Whitford, P.B. (1949). Distribution of woody plants in relation to succession and clonal growth. *Ecology*, 30: 199-208.

Wilke, B., Bogenrieder, A. and Wilmanns, O. (1993). Differenzierete Streuverteilung in Walde, ihre Ursachen und Folgen. *Phytocoenologia*, 23: 129-155.

Woods, P. V., and Raison, R. J. (1982). An appraisal of techniques for the study of litter decomposition in *eucalypt* forests. *Aust. J. Ecol.*, 7: 215-225.

Wright, S.J. and Schaik, C. P. V. (1994). Light and phenology of tropical trees. *American Naturalist*, 143: 192-199.

Wright, S.J., Duber, H.C. (2001). Poachers and forest fragmentation alter seed dispersal, seed survival, and seedling recruitment in the palm Attalea butyraceae, with implications for tropical tree diversity. *Biotropica*, 33: 583–595.

Xiong, S. and Nilsson, C. (1997). Dynamics of leaf litter accumulation and its effects on riparian vegetation: a review. *The Botanical Review*, 63: 240-264.

Zak, D. R., Holmes, W. E., Finzi, A. C., Norby, R. J. and Schlesinger, W. H. (2003). Soil nitrogen cycling underelevated carbon dioxide: a synthesis of forest face experiments. *Ecological Applications*, 13:1508-15.

Index

A

Abiotic environmental factors 75
Abiotic factors 75
Acacia 15
Acanthaceae 34
Adina cordifolia 12, 18, 24, 26, 27
Aerobic digestion 68
Ailanthus excelsa 14
Akhand Shikar' (tribal hunt) 5
Alangium lamarckii 12
Alstonia scholaris 14
Amazonian Equator 35
Amoora rohituka 14
Ampelocissus 12
Analysis of variance (ANOVA) 66
Andropogon ascinodes 16
Annual litter fall 61
Anogeissus latifolia 15, 18, 24, 26, 27, 53
Anthistiria gigantia 12, 16
Anthocephalus 14
Anthocephalus cadamba 27
Anthus rufulus 104
Aonyx cinerea 130
Apluda mutica 16, 17
Aquatic ecosystems 95
Arboreal animals 53
Arbortristis 15
Ardisia solanacea 12
Artocarpas lacoocha 14
Artraxon 16
Arundinella setosa 16
Arundo donax 17
Asparagus 13
Asparagus racemosus 154
Asteraceae 34
Asynchronous 80
Atmospheric CO_2 32
Available phosphorus 75
Avian conservation 102
Avifauna 102
Aythya ferina 104
Aythya fuligula 104

B

Badampahar reserve forest 18
Bahunia vahlii 12
Bamboo Pit Viper 101
Banded quartzites 10
Banded Racer 100
Baned Krait 101
Barehipani waterfall 3
Bark gecko 99
Barking deer 154
Barleria strigosa 12
Barred Wolf Snake 101
Basal area 70
Bassia 13
Bassia latifolia 18
Bataguridae 99
Bauhina vahlli 15, 16, 18
Bay of Bengal 8
Beaked Worm Snake 101
Bear 154
Beddome's Grass Skink 100
Binocellate Cobra 101

Biodiversity 2
Biodiversity assessment 128
Biodiversity conservation 95
Biodiversity monitoring programs 127
Biogeographic regions 1
Biological biodiversity 102
Biological diversity 1
Biosphere Committee 2
Biosphere reserve 54
Biosphere reserves 1
Biotic surveys 95
Bio-volume 70
Bird diversity 4
Biscophia javanica 14
Boerhavia diffusa 154
Bombax ceiba 12, 14, 24, 26, 27
Bore climbers 34
Botanical Survey of India 11
Bothriochloa bladhii 17
Bothriochloa sp. 16
Bridelia retusa 12, 14, 27
Buchanania lanzan 12, 16, 18, 15, 24, 152
Buffer areas 65
Buffstriped Keelback 100
Bulk density 71
Butea superba 16

C

Caesalpinaceae 34
Cage birds 105
Calcareous deposits 9
Camera traps 129
Canis aureus 130
Canopy 70
Cantor's Black-headed Snake 101
*Cappilipedium assimil*e 16
Carbon sequestration 60, 75
Cardiovascular diseases 154
Careya arborea 12, 16
Carissa spinarum 16, 18
Carnivores 103
Cassia fistula 12, 15, 16, 18
Cation exchange capacity 154
Cedrela ciliate 14
Cedrela toona 13, 16
Chahala 67
Checkered Keelback Water Snake 101
Chrysopogon aciculatus 16
Chrysopogon sp., 16
Chrysopogon verticillatus 16
Cissus 12
Clayey loam 9
Cleistanthus collinus 18
Climate 61
Climate change 61
Climatic factors 94
Climax vegetation 4
Cochlospermum religiosum 15
Colebrookia oppositifolia 16
Colossal sum 58
Combretaceae 34
Combretum decandrum 13, 18
Common Asian Toad 98
Common Blind Snake 101
Common Indian Bronze-back 100
Common Indian Cat Snake 100
Common Indian Krait 101
Common Indian Monitor Lizard 100
Common Indian Rat Snake 101
Common Indian Skink 100
Common Indian Tree frog 98
Common Indian Trinket Snake 100
Common Kukri Snake 101
Common Sand Boa 100
Common Snake Skink 100
Common Vine Snake 100
Common Wolf Snake 101
Communal hunting 150
Community structure 102
Complex organic constituents 61
Components of ecosystems 1
Compositional dynamics 21
Concentration of dominance (CD) 56
Conservation 131
Copper-Headed Trinket Snake 100
Coppice system 26
Croton roxburghii 12
Croton spp. 18
Curcuma 12
Curcuma aromatica 14
Cymbopogon flexuosus 16

Cymbopogon fresuosus 17
Cynodon dactylon 16, 17
Cyperaceae 34

D

Dalbergia sisoo 26, 27, 30
Deforestation 58, 95, 136
Delay forest recovery 34
Dendrophthoe falcate 152
Digitaria spp. 17
Dillenia pentagyna 12, 16
Dioscorea 12, 13
Dioscorea sps. 154
Diospyros melanoxylon 18, 24
Diospyrus embryopteris 14
Diospyrus melanoxylon 15
Dipterous 75
Dispersing seeds 127
Distribution of Climbers 52
Diversity of ground flora 93
Dobson's Burrowing Frog 98
Domestic Livestock Grazing 148
Dominant tree species 56
Double exponential model 61
Dry deciduous 2
Dry Deciduous Hill Forests 14, 19
Dry Sal Forests 19
Dry weather conditions 151
Dubois's Tree Frog 98
Dutta's Cricket Frog 98

E

Eastern Black Turtle 99
Eastern Bronze Skink 100
East Indian leopard gecko 99
Ecological amplitude 53
Ecological community 53
Ecological Dynamics 21
Ecological economics 77
Ecological information 13
Ecological life 93
Ecological rehabilitation 58
Ecological research 32
Ecosystem Service 59
Efficient utilization of pollinators 79
Elephant 154
Elongated Tortoise 99
Emblica officinalis 16
Enchytraeids 75
Endangered plants 154
Endangered species 131
Equilibrium litter 61
Eragrostis atrovirens 17
Eragrostis sp. 16
Erythrina suberosa 15
Essential oil 140
Eupatorium odoratum 22
Euphorbiaceae 34
Evapotranspiration 69
Evergreen forest 154
Evergreen species 92
Excelsum 12
Extinction of species 140

F

Fabaceae 34
Faunal diversity 3
Faunal Monitoring project 96
Felling operations 28
Ferguson's Toad 98
Ficus benghalensis 24
Ficus spp. 14
Fire season 150
Firewood 137
Flacourtia ramontchi 15
Flemingia 12
Flemingia chapper 13
Floral wealth 33
Floristic composition 33, 34
Floristic composition, physiognomy 12
Flowering activity 91
Footprints measurements 127
Forest biomass 54
Forest community 79
Forest Dynamics 19
Forest ecosystem 1, 136
Forest floor 22, 61
Forest floor biomass 62, 75
Forest floor litter 59
Forest floor microorganisms 62

Forest fragmentation 58
Forest lands 142, 148
Forest succession 53
Forsten's Cat Snake 100
Fragmentation 3
Frugivores 103
Fruiting activity 92
Fruit maturation 92
Fruit maturation activity 80
Fuel-wood 137
Fungoid Frog 98

G

Gallinago gallinago 104
Gardenia gummifera 17, 18
Gardenia latifolia 15, 18
Garuga pinnata 15
Gastrointestinal disorder 154
Geographic distribution 95
Giant squirrel 154
Girth classes 30
Girth limit 25
Globally threatened species 104
Glochidion 14
Gmelina arborea 16, 24, 26, 27, 30
Gondwana 8
Grainivores 103
Granitoid 9
Grassland and Savannah 16,19
Green forest canopy 154
Green leaf manure 76
Grewa sp. 12
Grewia 16
Grey Ballon Frog 98
Ground flora 24, 149
Growth dynamics 30
Gurandi 154
Gurguria 67
Gynecological disorder 154

H

Haematitic rock 9
Halcyon smyrnensis 104
Handipuhan 67
Helicteres isora 12
Herbaceous species 93
Herbivore 3
Herpetofauna 96
Herpetofauna conservation programs 96
Herpetofaunal Diversity 95
Herpetological inventories 96
Heteropogon contortus 16, 17
High level Sal Forests 19
High Level Sal Forests 15
Hirundo daurica 104
Holarrhena antidys enterica 18
Holarrhena antidysenterica 16
Holoptelia integrifolia 24
Homogeneous substrate 61
Human population dynamics 142
Humid tropical forests 75
Hymenodictyon 12

I

Illegal cutting of Trees 137
illegal smuggling of timber 139
Imperata arundinacea 12, 13
Imperata arundinaceae 16
Imperata cylindrica 16, 17
Importance value Index (IVI) 53
Importance Value Index (IVI) 34
Indian Bull Frog 98
Indian Burrowing Frog 98
Indian Chamaeleon 99
Indian Flapshell Turtle 99
Indian Garden Lizard 99
Indian Green Keelback 101
Indian house gecko 99
Indian Rock Python 100
Indian Roofed Turtle 99
Indian Skipper Frog 98
Indian tropical forests 54
Indigofera cassioide 12
Indigofera pulchela 16
Indigofera pulchella 13
Insectivores 103
In-situ conservation 1
Interphenophase durations 92
Intrinsic factors 93

Iseilema laxum 16
IUCN Red List of threatened species 130

J

Jenabil 67
Jerdon's Bull Frog 98
Jhoom 18
John's Sand Boa 100
Joranda 67
Jungle fowl 105

K

Kalikaprasad 67
Karanjia division 17
Keystone species 11
Khadias 10
King Cobra 101
Kydia calycina 12

L

Lagerstroemia parviflora 12
Langur 154
Lannea coromondelica 15
Lannea grandis 18
Lantena camera 21
Law enforcement 131
Leaf drop 80
Leaf flushing 91
Leaf litter 151
Leaf senescence 93
Leaf shedding 80
Leea 12
Leopard 154
Lignin and tannins 61
Limbless Skink 100
Limestone 9
Linear regression analysis 71
Litsea nitida 14
Litsea monopetala 142, 154
Litsea nitida 14
Litter bag technique 60
Litter decompositions 59, 77
Litter Depth 61
Litter dynamics 76
Local community 131
Long Term Research Network (LTRN) 58
Low level Sal Forest 16
Low Level Sal Forests 19
Lulung 67
Lutra perspicillata 130
Luxuriant vegetation cover 8

M

Macaranga peltata 14
Madhuca indica 24, 26, 27
Madhuca latifolia 18, 152
Magniferra indica 14
Mallotus philippensis 15
Malvaceae 34
Mammal diversity 4
Man and Biosphere (MAB) 6
Man and Biosphere programme 3
Mangifera indica 13, 26, 27
Mankidias 10
Marbled Ballon Frog 98
Marbled Toad 98
Mathematical indices 55
Melastomataceae 34
Micaceous schists 9
Micaschists 9
Michelia champaca 12, 13, 14, 24, 26, 27
Micro habitats 3
Millettia extensa 12
Mimosaceae 34
Mitragyna parviflora 12, 24, 26, 27
Mock Viper 101
Mohua flowers 149
Moist deciduous 2
Monocellate Cobra 101
Moraceae 34
Mugger Crocodile 99

N

National forest policy 33
Natural forest 136
Natural grandeur 7
Natural regeneration 28

Natural regeneration 17
Neotropical forests 35
Neyraudia reynaudiana 22
Nigirdha 67
Nitrate nitrogen 75
Nitrogen mineralization 3
Northern Tropical Moist Deciduous Forest 19
Northern Tropical Semi-evergreen Forest 14
Northern Tropical Semi Evergreen Forest 19
Number of trap nights 129
Nutrient cycling 59
Nutritional balance 3
Nyctanthes 15

O

Olive Keelback Water Snake 100
Omnivores 103
Onvention of International Trade in Endangered Species of Wild Fauna and Flora (CITES). 97
Organic carbon 152
Ornate Flying Snake 101
Ornate Narrow mouthed Frog 98
Ornithological history 103
Ougeinia oogeinensis 26, 27
Ougeinia sp. 16
Ougenia oogeinensis 27
Overstorey 80
Overstorey species 80

P

Painted Balloon Frog 98
Palaeozoic era 8
Palpala 8
Parthenium 21
Peacock 154
Peninsular Tent Turtle 99
Perennial streams 7
Phenological events 78
Phenological observations 79
Phenological patterns 94
Phenology 4
Phenology 78
Phoenix acaulis 16
Photoperiod 91
Photosynthesis 153
Phyllanthus emblica 18
Phyllites 9
Physico-chemical 13
Phytomass 70
Phytosociology 4
Phytosociology 32
Piscivores 103
Plain Sal Forest 17, 19
Plant community 21
Plant diversity 32
Plant physiognomic groups 70
Poaceae 34
Podadiha 67
Podiceps cristatus 104
Pogonatherum paneceum 17
Pollinating insects 92
Pollinating plants 127
Pollinidium angustifolium 17
Polyalthia 12
Polyalthia serasioides 14
Population dynamics 102
Potential pollinators 102
Predator and prey 127
Prerocarpus marsupium 18
Prey populations 3
Project Tiger 6
Protected area 137
Protected forests 11
Protium serratum 15, 27
Pterocarpus marsupium 12, 16, 18, 24, 26, 27, 30

Q

Quartzite hamatite's 9
Quartzites 9

R

RAI 129
Rainfall patterns 61
Rainy season 93
Randia 15

Random distribution pattern 35
Rapid fruiting activity 92
Ratufa indica 130
Raunkiaer's classification 35
Rauvolfia serpentina 154
Recurring biological events 78
Red-necked Keelback 100
Relative Abundance Index (RAI) 132
Relative abundance of mammals 129
Renewable genetic resources 32
Reptilian conservation 97
Resins 140
Retusa, Bridelia 27
River Khairi 8
Rubiaceae 34
Russell's Viper 101
Rutaceae 34

S

Sacciolepsis indica 17
Sal forests 138
Salix tetrasperma 14
Sapling layer 55
Saraca ashoka 154
Saw-scaled Viper 101
Schleichera oleosa 24
Seed dispersal 79
Seed dispersers 102
Seedling layer 55
Seedlings 62
Sehima nervosa 17
Semicarpus anacardium 15
Semi-evergreen 13
Shales 9
Shannon Wiener's index 55
Shorea robusta 11, 17, 22, 26, 27, 30, 53
Short-headed Burrowing Frog 98
Shrub savannah 152
Silvicultural system 12
Similipal Biosphere Reserve Bush Frog 98
Similipal Biosphere Reserve (SBR) 34
Similipal forests 5
Similipal Tiger Reserve (STR) 6
Single exponential model 61
Skeletal diseases 154
Skin diseases 154
Smilax 12
Smilax macrophylla 12, 154
Smilax sp. 18
Smilax zeylanica 16
Smooth Water Snake 100
Snake-eyed Lacerta 100
Soil biota 61
Soil biota 75
Soil chemical characters 75
Soil fauna 75
Soil fertility 60
Soil moisture content 71
Soil nutrient status 152
Soil organic carbon 75
Soil organic pool 59
Sono 8
Spatial distribution of plants 148
Species diversity 55, 56, 70
Species dynamics 22
Species evenness (SE) 56
Species richness (SR) 56
Sporobolus indicus 17
Spotted deer 154
Spotted Indian house gecko 99
Springtails 75
Sterculia urens 15
Sterculia villosa 15
Sterospermum suaveolens 14
Strobelanthus 13
Structural dynamics 21
Structural Dynamics 23
Subarnarekha river 8
Submetamorphic 9
Succession 21
Succulent grass 150
Sus scrofa 130
Sustainable resource management 58
Syhadra Cricket Frog 98
Symbopogon martini 12
Symplocos racemosa 16
Symplocos sp. 17
Systematic exploitation 26
Systematic plan 30
Syzygium cerasoides 16

Syzygium cumini 14, 12, 24, 26, 27
Syzygium sp. 17

T

Terminalia alata 12, 15, 16, 18, 24, 26, 27, 53, 152
Terminalia arjuna 14
Terminalia belerica 24
Terminalia chebula 15, 16, 24
Terrestrial mammals 127
Tetracerus quadricornis 130
Themeda 16
Themeda arundinacea 16
Themeda caudate 16
Themeda quadrivalvis 16
Themeda spp. 17
Themeda triandra 16
Themeda villosa 16
Threatened species 104
Threatened vertebrate animal 95
Thunderstorms 8
Thysanolaena maxima 22
Tiger 154
Toona ciliata 14
Topography 7, 61
Total forest floor material 59
Tree density 70
Tree Felling 27
Tree mortality 23
Tree savannah 152
Trema orientalis 13
Trewia nudiflora 14
Tribal forest dwellers 142
Tricarinate Hill Turtle 99
Tropical forests 136
True genesis 9
Turtles (soft-shell) 97
Twin-spotted Wolf Snake 101
Typical savannah 149

U

Unclassified forests 11
Understorey 80
Understorey species 80
United Nations Educational, Scientific, and Cultural Organization (UNESCO) 1
Upland ecosystems 76

V

Varzea forest 35
Vegetational layers 55
Vegetation dynamics 62
Vegetation layers 52
Ventilago denticula 15
Vulpes bengalensis 130

W

Water stress 92
Wendlandia 12
Wendlandia sp. 53
Wendlandia tinctoria 16, 18
White-Spotted Supple Skink 100
Wild boar 154
Wild cat 154
Wild Fire 149
Wildlife 5
Wildlife habitats 33
Wildlife monitoring 128
Wildlife (Protection) Act 97
Wildlife sanctuary 10
Wildlife Wing 97
Wild pig 154
Woodland species 104
World Conservation Monitoring Centre [WCMC] 32

X

Xylia xylocarpa 12, 14, 24, 26, 27

Y

Yellow Monitor Lizard 100

Z

Ziziphus 15
Ziziphus sp. 18

www.ingramcontent.com/pod-product-compliance
Ingram Content Group UK Ltd.
Pitfield, Milton Keynes, MK11 3LW, UK
UKHW021950270726
14060UKWH00002B/443